WIMBLEDON PUBLISHING C...
LIBRARY OF QUALITY IN EDUCATION

# CHEMISTRY
## For Key Stage 3

---

By
RICHARD BALDING
Summer Fields, Oxford

GENERAL EDITOR
K.S. Sood
LIBRARY OF QUALITY IN EDUCATION

*Wimbledon Publishing Company*

ISBN 1 898855 18 8
*Dedicated to the Very Highest Quality in Education*

# Introduction

## TO THE STUDENT

Acid Rain, the ozone layer, limescale in kettles and water pipes, the burning of hydrocarbons ...

These are all today's issues and require a simple knowledge of chemistry in order to understand them.

However, many students find that chemistry involves learning a great deal of names and, although the experiments are fun, it is not always easy to relate the correct substances to any particular experiment!

This book has been written to help you to understand the basic principles of chemistry and to place a few of the substances which you come across at this stage, into their rightful places.
The material in the book covers the factual material which is required by the National Curriculum at Key Stage 3 and contains material which will help you in your preparation for Common Entrance and Public Schools Scholarship Examinations, as well as providing a firm base for your first year of GCSE work.

There are two parts to each chapter: the **text**, which sets out the material which you are expected to know and **questions** for you to answer to make sure that you have understood the principles. Scholarship Examinations are designed to find out how well you can apply basic principles to solve problems which may at first appear unfamiliar or more complex, i.e. they are really testing your ability to **think**.

Remember, at this level if will be assumed that you *know* the basic facts of the subject and it is how you can *apply* these facts to produce a clear, concise and well-presented answer to problems which will determine how successful you are.

The sets of 'Further Questions' are designed to give you practice at scholarship level.

It may well be that you have missed some of the topics at some time during your Science Course, or there may have been some areas in which you did not understand fully the basic principles involved and thus found your studies less rewarding than they might have been. This book is therefore designed as a 'revision aid' and/or a 'gap filler'.

# TO THE TEACHER

This book covers material required to satisfy the requirements of the National Curriculum, Key Stage 3, Attainment Target 3 and it is hoped that it will be a useful addition to course material already used. The book is designed to help you in the following ways:

- ☐ as a check list of the important topics which need to be covered at this stage;
- ☐ as a supply of questions for reinforcement, assessment and homework purposes;
- ☐ as a source of revision material prior to examinations;
- ☐ as self-help material which can be given to a student who is away from the laboratory for any length of time (say for illness, or for holiday work).

As this book is designed to help students increase their knowledge and understanding of the required factual material, practical work (such as that detailed in Attainment Target 1) has not been included, even though it is a vital part of any science course. All details of experiments and/or demonstrations have been included only as reinforcement of work which you may well have done in class time.

**It is important to realise that this book is an extra resource to *complement*, rather than *replace*, existing materials which will, of course, include details of practical and investigative work.**

# Contents

## 2 "MAKE IT PURE" — SEPARATING MIXTURES (continued)



## 3 HEAT IT AND CHANGE IT



## 4 INDICATORS AND THEIR USE IN CHEMISTRY

## 4 INDICATORS AND THEIR USE IN CHEMISTRY (continued)



## 5 GASES – IN GENERAL AND IN PARTICULAR

## 5 GASES – IN GENERAL AND IN PARTICULAR (continued)



## 6 METALS AND NON-METALS

## 7 COMPARING THE REACTIVITY OF SOME ELEMENTS



## 8 METALS AND MINERALS FROM ROCKS

## 8 METALS AND MINERALS FROM ROCKS (continued)



## 9 ELECTRICITY IN CHEMISTRY — AN INTRODUCTION

# "SUGAR AND SPICE AND ALL THINGS NICE" — a first look at matter, elements, compounds and mixtures

# 1

WHEN YOU THINK OF CHEMISTRY, do you think of test tubes, protective eye-glasses and Bunsen burners? Yes, these *are* all parts of your chemistry course, but there is much more to chemistry.

The shampoo which you use to wash your hair will have been formulated by a chemist.

The dyes which colour your clothes will have been tested by a chemist to make sure that not only are they suitable for material but also that the colour will not come off the material onto you.

A chemist who is also interested in biology and medicine (often called a biochemist), will have worked out which chemicals to put into the pills which we take when we have a headache.

In short, chemistry is **all around** us and it is the study of the materials which make up the world in which we live. Everything, including the air, land, sea and all living organisms are made from a variety of substances to which we give the collective name, **matter**.

## What is matter?

All substances are made from tiny particles called **atoms**. There are about 100 different type of atoms. The word 'atom' comes from the Greek word, *atomos* (which means indivisible) and atoms are so called, because they are the **smallest** particle which can exist.

Atoms join together to form **molecules**.

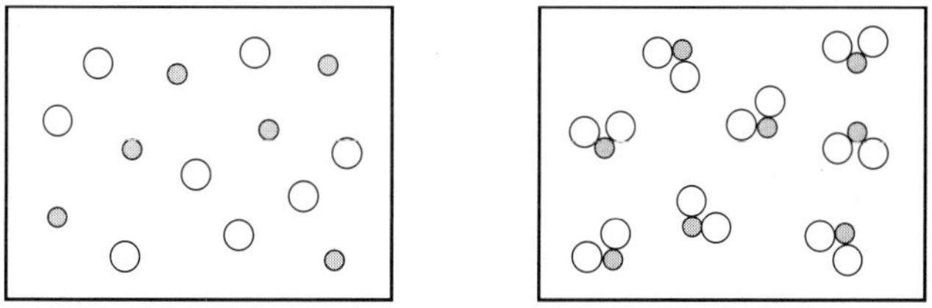

Fig. 1.1 *Atoms and molecules*

If you look at the variety of materials around you, you will see that most of them can be put within three main groups: SOLIDS, LIQUIDS or GASES. These groups are known as the **states of matter**.

A given amount (mass) of a substance in each of the three states will show the following features (**properties**):

**TABLE 1.1 STATES OF MATTER**

| Solid | Liquid | Gas |
|---|---|---|
| Fixed Mass | Fixed Mass | Fixed Mass |
| Fixed Volume | Fixed Volume | Changes Volume |
| Fixed Shape | Changes Shape | Changes Shape |

Liquid and gases are called **fluids** because they can change shape and flow.

## How do substances change shape?

All molecules are **constantly moving** – we say that they have kinetic energy. Sometimes this movement is from one place to another as in fluids. At other times, the movement takes the form of vibrations in **one** place – as in solids.

As molecules come closer together – of the order of millionths of the thickness of hair – there are forces of attraction which bind them together.

There are various types of forces and methods of holding molecules together, but we shall simply group these together and call them **binding energy**.

It is important to realise that:

(i) EACH substance has **ITS OWN** binding energy, which does not change.
(ii) The kinetic energy of the molecules will increase when substances are heated.
(iii) In **SOLIDS**, the kinetic energy of the molecules is LESS than the binding energy – so solids retain their **shape**.
(iv) In **LIQUIDS**, the kinetic energy of the molecules is ALMOST EQUAL to the binding energy – so molecules are able to move around but are generally still held within the liquid. In this way the liquid is able to change shape although it still keeps the same volume.

(v) In **GASES**, the kinetic energy of the molecules is GREATER than the binding energy – so the molecules are free to move, if they can.

Study Fig. 1.2 below to make sure that you understand the relationship between kinetic energy and binding energy in solids, liquids and gases.

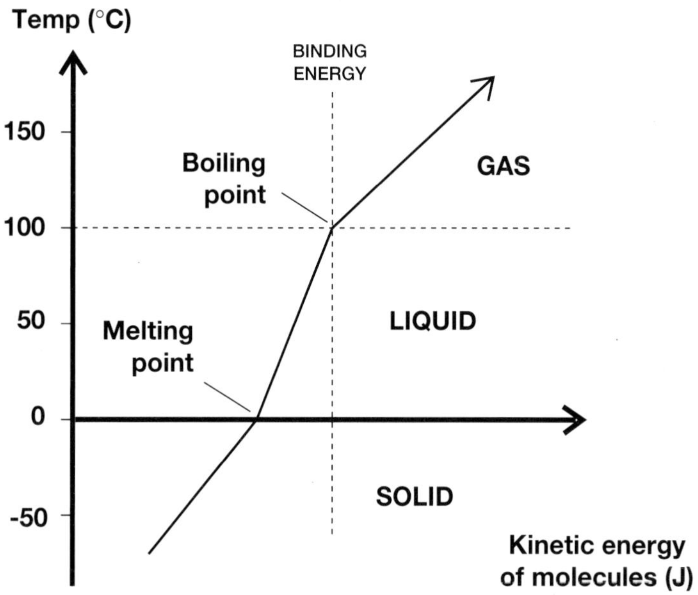


Fig. 1.2 *The relationship between temperature, kinetic energy of molecules and binding energy of a substance in solids, liquids and gases*

## How do substances change shape?

When we heat substances, we add energy to them. This extra energy is added to the already-moving molecules and so they move faster.

When we cool a substance, energy is transferred **away** from the molecules and so they, now having less energy, slow down.

You will be familiar with the three states of water.

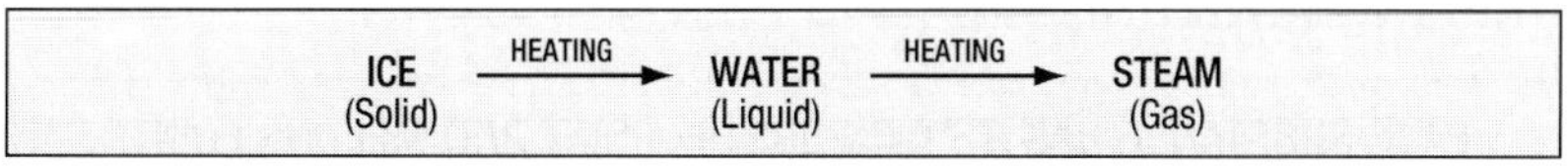


(See Fig. 1.3 for more detail.)

### ICE (SOLID)

The kinetic energy of the molecules is LESS than the binding energy, so the molecules are held in place with respect to each other. So the solid ice retains its shape.

### WATER (LIQUID)

The molecules have more kinetic energy and so move further apart. However, the amount of kinetic energy which each molecule has is **slightly less** than the binding energy of water. This means that although the molecules are free to change places with each other, most of the molecules do not have enough energy to break free of the liquid.

### STEAM (GAS)

Here, the molecules have more kinetic energy than the binding energy of water and are able to break free of the liquid. Because the molecules of a gas are free to roam, the gas will completely fill any container.

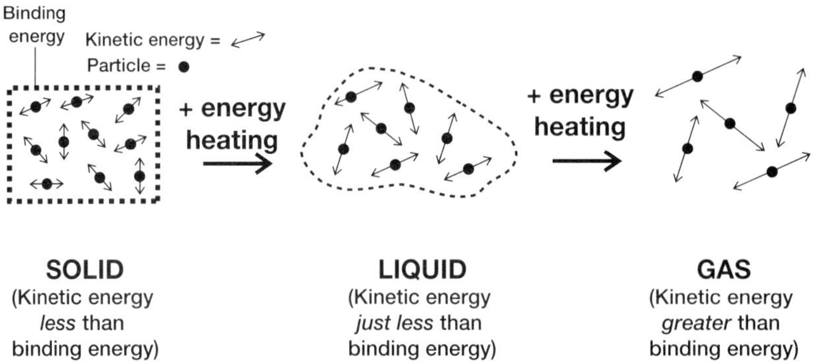


Fig. 1.3 *How molecules behave as solids change into liquids and gases*

## Boiling point and melting point

The temperature at which a liquid changes into a gas is called the **BOILING POINT**.

The temperature at which a liquid changes into a solid, is called the **MELTING POINT** (or FREEZING POINT).

| EACH SUBSTANCE HAS ITS OWN BOILING POINT AND MELTING POINT* |
|---|

* This is true **ONLY** at a given external pressure – often referred to as **standard pressure**. Chemists often refer to them as **fixed points**.

## USING BOILING POINT AND MELTING POINT TO DETERMINE IF A SUBSTANCE IS PURE

The fact that **each** substance has its **own** boiling point (and melting point) can be used to help identify a substance. For example, consider two colourless liquids below:

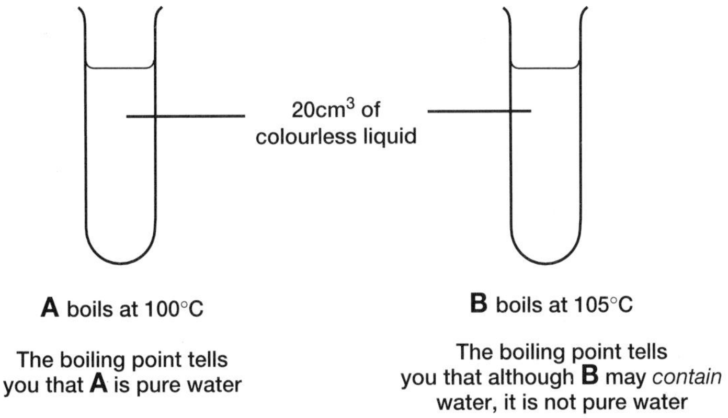


Fig. 1.4

**NOTE**

**IMPURITIES RAISE THE BOILING POINTS AND LOWER THE FREEZING POINTS OF SUBSTANCES**

The process of changing a liquid into a gas is called **EVAPORATION**.

The process of changing a gas into a liquid is called **CONDENSATION**.

BOTH of these processes take place at the **boiling point** of the substance.

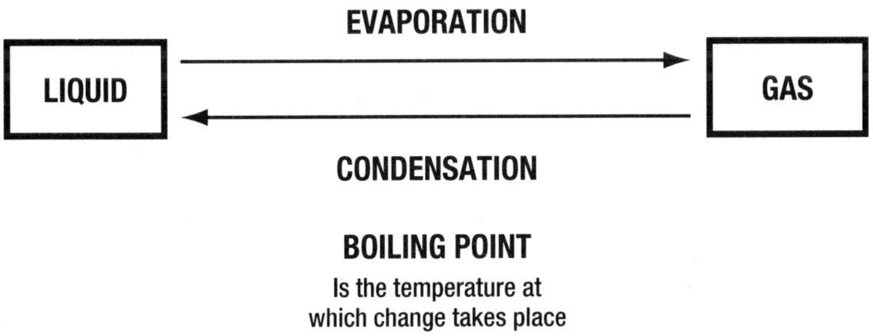

**REMEMBER**

**EVAPORATION AND CONDENSATION ARE PHYSICAL CHANGES**
NO CHEMICAL REACTIONS ARE INVOLVED
THERE IS NO CHANGE IN THE CHEMICAL COMPOSITION

## Sublimation

Generally, when solids are heated, they may change into a liquid state and then into a gas. Likewise cooling a gas will change it into a liquid and on further cooling, into a solid.

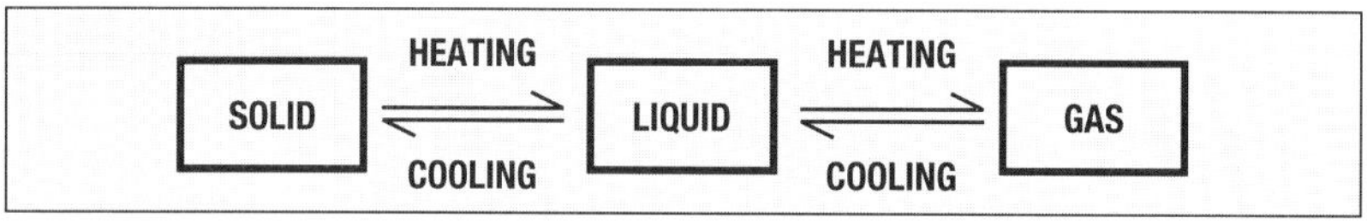


However there are some substances which 'miss out' the liquid state and change directly from solid to gas and from gas to solid. The process of missing out the liquid state is known as **SUBLIMATION**.

IODINE, and CARBON DIOXIDE are examples of substances which SUBLIME.

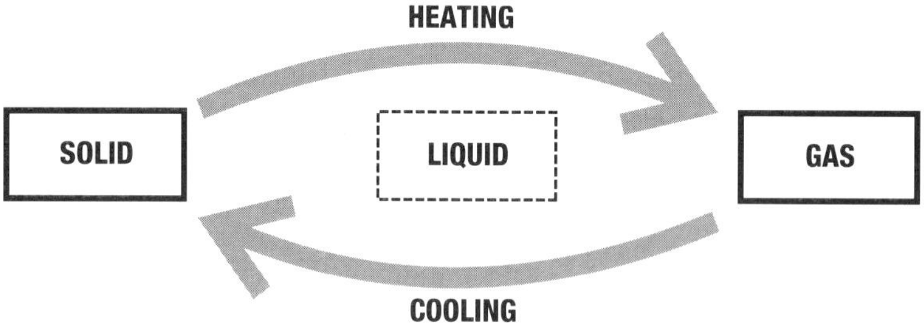


Fig. 1.5 *Sublimation*

**FOR EXAMPLE**

(i) **IODINE (s)** (BLACK SHINY SOLID) ⇌ (SUBLIMATION) **IODINE (g)** (PURPLE GAS)

(ii) **CARBON DIOXIDE (s)** (WHITE SOLID) ⇌ (SUBLIMATION) **CARBON DIOXIDE (g)** (COLOURLESS GAS)

Boiling point and melting point are examples of **PHYSICAL PROPERTIES** (features) of a substance.

## What are physical properties?

These are properties (features) which are special to substances and which may be observed and/or measured, **without the substances changing into another substance**.

### SOME MORE PHYSICAL PROPERTIES

**1. Density**

This is an important physical property of a substance for, like boiling point/melting point, we can use it to identify a substance.

Each material has its own special number called its DENSITY, which describes the amount of matter which is padded into a specified volume of it.

The specified volume is usually 1 $m^3$.

The **AMOUNT** of matter in a given quantity of material is its **MASS** – measured in **kg**.

The **QUANTITY** of the material is its **VOLUME** – measured in $\mathbf{m^3}$.

The identity of a material is the mass present **in each** $m^3$ and can be found by dividing the total mass of the material by the volume of it.

So,

$$\text{Density}\left(\frac{\text{kg}}{\text{m}^3}\right) = \frac{\text{Total MASS}}{\text{Total VOLUME}}\left(\frac{\text{kg}}{\text{m}^3}\right)$$

In your laboratory work, you will probably have measured mass in g and volume in $cm^3$, rather than kg and $m^3$. This is fine and we shall express density as the mass (g) of each $cm^3$ of material.

**TABLE 1.2 SOME DENSITIES OF SUBSTANCES (AT STANDARD PRESSURE AND 20°C)**

| **Solids** | | **Liquids** | | **Gases** | |
|---|---|---|---|---|---|
| Lead | 11.3 g/cm$^3$ | Mercury | 13.6 g/cm$^3$ | Chlorine | 0.0015 g/cm$^3$ |
| Copper | 8.9 g/cm$^3$ | Water | 1.0 g/cm$^3$ | Oxygen | 0.0013 g/cm$^3$ |
| Iron | 7.9 g/cm$^3$ | Ethanol | 0.8 g/cm$^3$ | Nitrogen | 0.0009 g/cm$^3$ |
| Aluminium | 2.7 g/cm$^3$ | | | Hydrogen | 0.00007 g/cm$^3$ |

You are **not** expected to remember the densities of substances, but as you progress through your course, you will use the density of water so many times that you will remember it.

**2. Strength**

This can be divided into two types:

(i) **OVERALL STRENGTH** – Ability to withstand loads without deforming or breaking
(ii) **TENSILE STRENGTH** – Ability to change shape without breaking, e.g **FLEXIBILITY** – can be bent or twisted.
**MALLEABILITY** – can be hammered into shape
**DUCTILITY** – Can be drawn into wires

**3. Hardness**

One of the simplest and most often used tests for hardness is a Scratch Test.

If *x* makes a scratch **on** *y*, then *x* is **harder** than *y*.

**4. Conductivity**

Some materials will allow energy (thermal/electric) to pass through them. For example, electricity will pass through (i.e. is **CONDUCTED** by) wires made from metals such as copper.

Materials which do not allow energy to pass through them are called **INSULATORS**. Plastics are good insulators of energy.

As well as physical properties, a chemist will be interested in the **chemical properties** of materials.

## What are chemical properties?

These describe the composition of a substance and how, and under what conditions, it changes into another substance. When a substance changes to a completely different substance, then we say that a **chemical reaction** has taken place.

Clearly, we shall look further at chemical properties of substances as we progress through this book.

## Types of substances

In order to understand more about substances, we need to know about the different types of substances which make up the world around us. Substances may be found in their pure forms, or as part of a mixture:

(i) **Pure substances** – contain **one** substance only. This group is made up of **elements** and **compounds**.

(ii) **Impure substances** – contain two or more substances. Substances in this group are called **mixtures**.

## PURE SUBSTANCES – ELEMENTS AND COMPOUNDS

### Elements

There are about one hundred of these and they are all listed in the **periodic table** (we will discuss this further, below).

Elements are single substances which are the building blocks of all matter.

The smallest particle of an element which is recognisable as that element is an **atom** of that element. Although the word 'atom' means indivisible, we now know that an atom is made up of various parts, so we will look at the structure of atoms.

### Structure of atoms

All atoms are made up of tiny particles, some of which are electrically charged – the exact nature and structure of these tiny particles is still the subject of research. We do know, however, that each atom consists of a nucleus which is **positively** charged, surrounded by a 'cloud' of electrons which are negatively charged.

We also know that there are three main types of particle:

(i) **Proton** – a **positively** charged particle which is present in the nucleus.

(ii) **Neutron** – a particle which has mass, but has no electric charge. This particle is also found in the nucleus.

(iii) **Electron** – a particle which has a much smaller mass than that of a proton. An electron carries a **negative** charge and moves around the nucleus in a particular orbit which chemists call a shell. Different atoms have different numbers of shells and there are rules which determine how many electrons can be in any one shell.

The shells represent levels of energy which electrons in any particular shell possess and you will study this aspect of atomic structure in more detail later in your science studies.

We say that a nucleus is surrounded by a 'cloud' of electrons, because it is impossible to say exactly what the speed and location of an electron is, at any given moment. This was proved by a famous scientist called Heisenberg and is generally known as Heisenberg's Uncertainty Principle.

The charges are balanced i.e. all atoms are electrically neutral. This means that:

| THE NUMBER OF **PROTONS** = THE NUMBER OF **ELECTRONS** |
| --- |

Think about it. The chemicals in your laboratory are quite stable – there is no electrical conflict in a jar of iron filings!

Fig. 1.6 below gives the general structure of an atom.

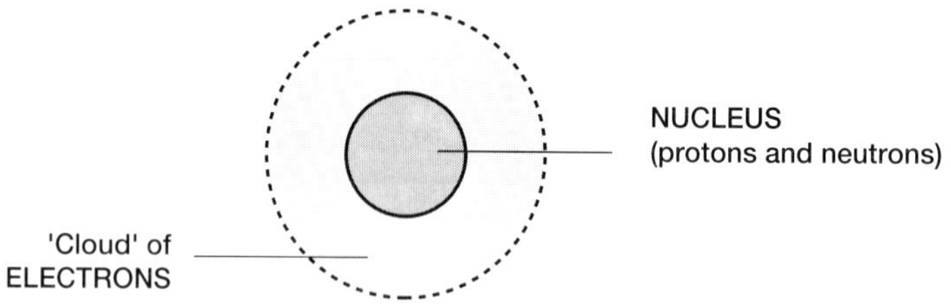


Fig. 1.6 *General structure of an atom*

Elements are, therefore, materials which are made up of **one type of atom** i.e. **all** the atoms of magnesium will be alike, in that they contain the same number of protons in their nuclei. In fact a magnesium atom has 12 protons in its nucleus.

Although oxygen atoms will contain the same basic particles (i.e. protons, electrons, neutrons) they will contain a different **number** of these to magnesium, so oxygen is a totally different material. In fact an oxygen atom has 8 protons in its nucleus.

The simplest atom of all is a hydrogen atom. This is simply 1 proton which is orbited by 1 electron.

*The periodic table*

This is basically a list of all the elements arranged vertically into 'families' or **groups**. The groups are numbered (usually with roman numerals) from I to O (also called group VIII).

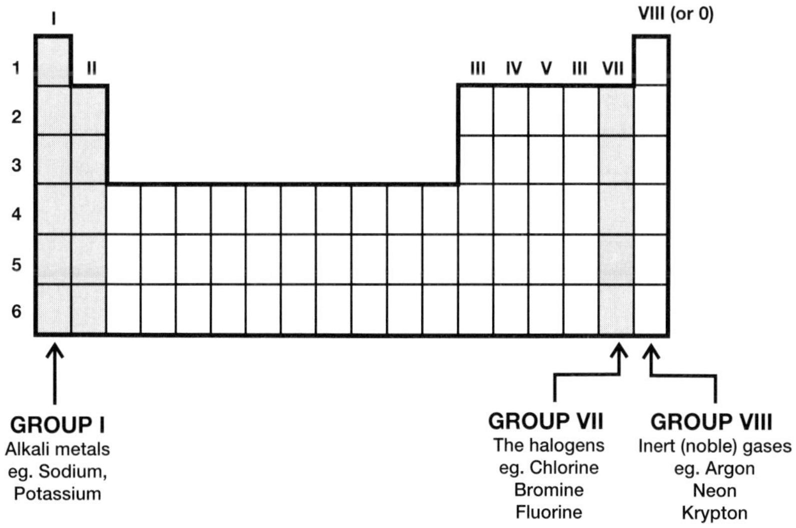


Fig. 1.7

The numbers on the left of the table denote the **period** the element is in.

The whole table was drawn up by a Russian Chemistry teacher called Dmitri Mendeléev in 1869. The brilliance of this table is that when Mendeléev drew this table up, not all the elements which are listed on it today had been discovered – so he left **gaps**. Because these gaps formed part of a 'family' of elements whose properties were known, he was able to predict the properties of yet undiscovered elements. Thus he was able to predict what Geranium would be like, even though this element was not discovered until seventeen years after Mendeléev had drawn up his table.

Later on in your chemistry studies, you will find the periodic table extremely useful, especially when you come to study and use atomic structure in more detail.

As an interesting pointer to those who wish to study atomic structure and the periodic table in more detail, it is useful to point out:

(i) The number of electrons in the **outermost** shell is given by the group number. It is the number of electrons in the outermost layer (or shell) which determines the chemical characteristics of that element.

(ii) The numbers of shells of electrons is given by the period number.

(iii) The chemical elements are usually listed on the periodic table using their chemical symbols.

*Chemical symbols*
This is a shorthand way of writing chemical elements. Each symbol is **always a capital letter** which is sometimes followed by a lower case one.

**FOR EXAMPLE**

C – Carbon
Cl – Chlorine
Co – Cobalt
Cs – Caesium

Sometimes the symbols are based on the Latin names of the elements.

**FOR EXAMPLE**

Pb – Plumbum – Lead
Cu – Cuprum – Copper
Na – Natrium – Sodium
K – Kalium – Potassium

You are not expected to remember, or use, chemical symbols at this stage. (Still, they are very useful, and you will see then often, particularly in the periodic table which is probably on the wall in your laboratory.

**Compounds**
When two, or more elements combine as a result of a **chemical reaction**, then a compound is formed.

COMPOUNDS ARE ENTIRELY DIFFERENT IN ALL RESPECTS FROM THE ELEMENTS WHICH REACTED TO FORM IT

For example when you burn magnesium ribbon in air, you are really enabling the oxygen in the air to combine with the magnesium ribbon. As you know, there is a white blinding flash as the reaction takes place and you are left with a white powder which is called magnesium oxide.

This magnesium oxide is a **compound** of the two elements – magnesium and oxygen.

We usually express the reaction in the form of a **word equation** with an arrow "→" which means "goes to form".

**FOR EXAMPLE**

MAGNESIUM (A silvery metal) + OXYGEN (A colourless gas) → MAGNESIUM OXIDE (A white powder)

**NOTE**

IN A COMPOUND, THE AMOUNTS OF THE COMBINED ELEMENTS ARE IN A FIXED RATIO TO EACH OTHER

COMPOUNDS CAN **ONLY** BE FORMED AND SPLIT UP (DECOMPOSED) BY MEANS OF CHEMICAL REACTION

When compounds are **formed** – we say that a COMBINING CHEMICAL REACTION has taken place.

When compounds are split up (decomposed) – we say that a **decomposing reaction** has taken place.

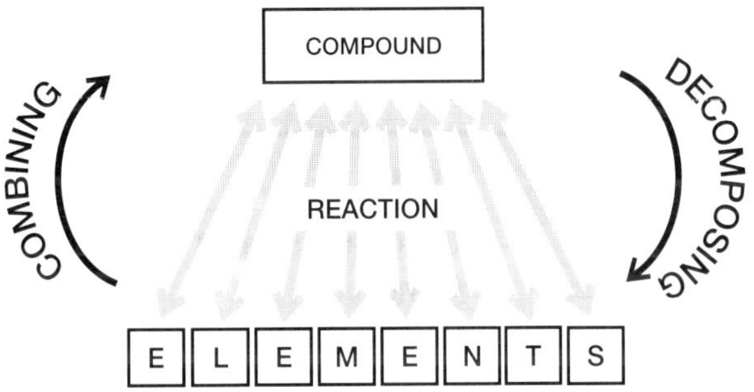


Fig. 1.8

## IMPURE SUBSTANCES – MIXTURES

When you go to the stock bottle to collect some hydrated copper sulphate, you would expect it to be pure, i.e. contain **only** blue crystals of hydrated copper sulphate. In other words, you expect the bottle to contain a **pure** substance, namely the compound hydrated copper sulphate.

The same could be said for bottles of sulphur, carbon, copper carbonate and so on.

# What do you know about ...

1. Explain what is meant by the terms:

   (a) atom
   (b) molecule

2. Matter exists in one of three states. Say what these states are and list three properties of each state.

3. Write down two things which happen to the molecules of a substance when that substance is heated.

4. Draw diagrams to show what happens to the molecules when ice becomes water, water becomes steam.

5. In each case, give the name of the process which best describes the changes below:

   (a) SOLID → LIQUID
   (b) SOLID → GAS
   (c) LIQUID → GAS
   (d) LIQUID → SOLID
   (e) GAS → LIQUID
   (f) GAS → SOLID

6. Explain the terms BOILING POINT and MELTING POINT and say why these are often called FIXED POINTS.

7. What is the difference between the melting point and freezing point of a pure substance?

8. The changes which are listed in question 5 are physical changes. What does the term 'physical change' mean? You should list at least two features.

9.

| Substance | A | B | C | D | E |
|---|---|---|---|---|---|
| Melting point (°C) | –73 | –120 | –7 | –182 | 649 |
| Boiling point (°C) | –10 | –196 | 59 | –164 | 1107 |

   (a) Which substance is a liquid at room temperature? Give a reason
   (b) Which solid is a solid at room temperature? Give a reason.
   (c) Which gas would be the most difficult to liquefy? Give a reason.

## ... matter, elements, mixtures and compounds?

# What do you know about ...

10. Carbon dioxide sublimes at –78°C. The *solid* form of carbon dioxide is called *dry ice*.

    (a) Draw a labelled diagram which includes all of the information given above.
    (b) Why do you think that solid carbon dioxide is called dry ice?

11. Pure water boils at 100°C and freezes at 0°C.

    (a) What would be the effect on the fixed points if salt was added to water?
    (b) Carefully explain why local councils often put salt on the roads during winter.

12. (i) What is meant by the density of a substance?
    (ii) Explain how a knowledge of density is useful to a chemist.

13. Here is a list of some densities:

| | |
|---|---|
| ALUMINIUM | 2.7 $g/cm^3$ |
| IRON | 7.8 $g/cm^3$ |
| LEAD | 11.3 $g/cm^3$ |
| GOLD | 19.3 $g/cm^3$ |

    (a) A metal block has a mass of 540 g and a volume of 200 $cm^3$. What might the block be made of?
    (b) I have been told that a box contains 1000 identical iron balls. The mass of one of the balls is 0.33 g and the volume of ONE of the balls 0.003 $cm^3$. Are the balls made of iron? If they are not, suggest a material from which they are made.

13 Say what the terms 'physical properties', and 'chemical properties' mean. Give examples of each and explain the differences between the two terms.

14. (a) What is an element?
    (b) Approximately, how many elements are there?
    (c) Where can you find a list of all known elements?
    (d) What can you say about atoms of elements? (Try to find two things.)

## ... matter, elements, mixtures and compounds?

# What do you know about ...

15. (a) What are, and where can you find:

    (i) protons
    (ii) neutrons
    (iii) electrons

    (b) Draw a labelled diagram of an atom to include the three items listed in part (a) of this question.

16. What can you say about the numbers of protons and electrons in an atom?

17. Who was Dmitri Mendeléev and why is his work so useful to chemists today?

18. Use the terms, *combining*, *decomposing*, *chemical reaction*, to describe the differences between an element and a compound.

19. Explain the difference between a compound and a mixture.

20. Can a mixture be a *pure* substance? Give a reason for your answer.

21. What is:

    (a) a solute;
    (b) a solvent

22 Explain, giving at least two examples, the difference between a solution and a suspension.

23. What is an emulsion? Give an example.

## ... matter, elements, mixtures and compounds?

# What do you know about …

*Use the information about water given below to answer questions 24, 25 and 26:*

Density of water at 0°C – 1.00 g/cm$^3$
Density of water at 20°C – 0.99 g/cm$^3$
Density of ice at 0°C – 0.92 g/cm$^3$
Melting point of ice – 0°C
To raise the temperature of 1 g of water by 1°C, requires 4.2 J of energy.
To melt 1 g of ice into water requires 330 J of energy.

24. How can you tell that ice floats on water?

25. If an ice cube has a mass of 10 g, calculate:

    (a) its volume;
    (b) how much energy is needed to melt it;
    (c) how much energy is required to heat the water formed from the melting ice by 10°C.

26. You have a drink which is 200 g of water at 20°C.

    (a) How much energy would need to be removed from the water to get its temperature down to 0°C?
    (b) Would a 10 g cube of ice be able to cool your drink to 0°C? Explain your answer.

## … matter, elements, mixtures and compounds?

# 2

# "MAKE IT PURE" — separating mixtures

First let us look at the main differences between compounds and mixtures.

| Compounds | Mixtures |
|---|---|
| Need a **combining chemical reaction** to form and a **decomposing chemical reaction** to split up | No chemical reactions involved in mixing and separating |
| Compounds are **entirely different** from the elements which make up the compound | Properties of the individual substances being mixed are retained |
| Elements are present in **fixed** amounts | Amounts of parts of a mixture can vary |

Thus, separating mixtures involve **PHYSICAL PROCESSES** and the component parts of a mixture are not changed in their composition as mixtures separate.

The key to separating mixtures is:

**FIND SOMETHING THAT ONE PART OF THE MIXTURE DOES, BUT THE OTHER DOESN'T**

**FOR EXAMPLE**

| MIXTURE | |
|---|---|
| 1. Sand + sawdust | Sawdust floats in water – sand doesn't. |
| 2. Sand + iron filings | Iron filings are attracted by a magnet – sand isn't. |
| 3. Sand + salt | Salt dissolves in water – sand doesn't. |

In (1), the separation uses the physical property of **density differences** between the two substances.

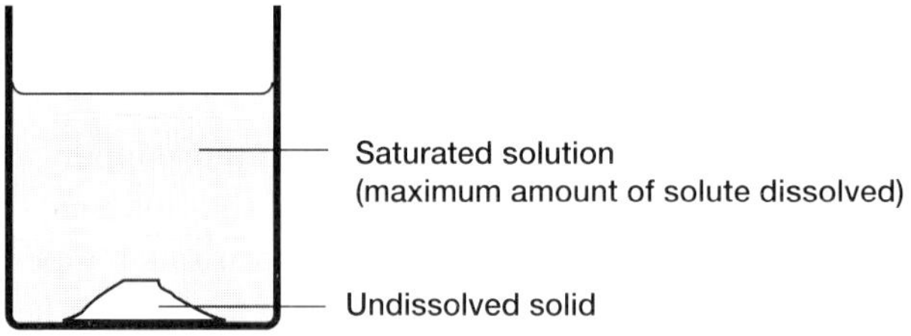


Fig. 2.2 *A saturated solution*

Think of a sponge. It will soak up water and become wet – however, it does not drip. There comes a time when it cannot absorb any more water i.e. it becomes **saturated**.

When it reaches this point, any more water added will not be absorbed by the sponge and so will drip off it.

## Crystallisation

If you have 100 g of a saturated solution of hydrated copper sulphate at 80°C then 58 g of solute will be dissolved. If this solution is allowed to cool, then the amount of copper sulphate dissolved becomes less and **crystals** of undissolved copper sulphate form on the bottom and sides of the beaker. As the solution cools, so more of the solid becomes undissolved and so the crystals increase in size.

If a '**seed**' crystal is suspended in a saturated solution which is allowed to cool, then the undissolved solid will form around the seed crystal making it larger. This is how you will have 'grown' large crystals of hydrated copper sulphate or 'Alum'.

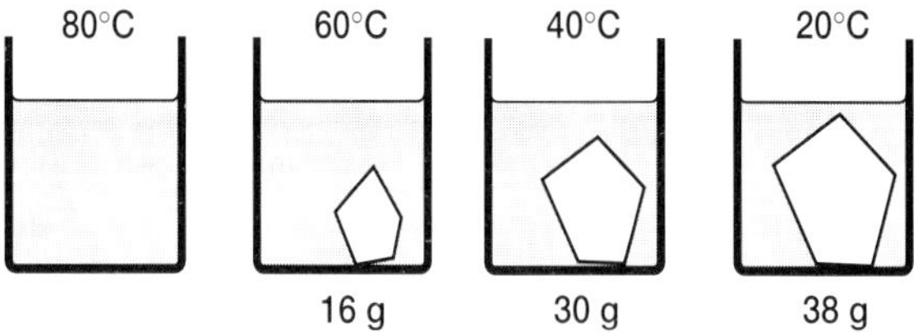


| | 80°C | 60°C | 40°C | 20°C |
|---|---|---|---|---|
| Amount dissolved | 58 g | 42 g | 28 g | 20 g |
| Amount undissolved | — | 16 g | 30 g | 38 g |
| TOTAL | 58 g | 58 g | 58 g | 58 g |

Fig. 2.3 *Crystallisation of hydrated copper sulphate in 100 g of water*

## Recovering the solute from a solution

There are two methods:

### 1. CRYSTALLISATION

As we have just seen, if a saturated solution is allowed to cool then crystals of **pure** solute may form. The surplus liquid is then poured away and the crystals can be dried and collected.

### 2. EVAPORATION

This is the **ONLY** way of recovering a solvent from a solution. As a solution is heated the solvent, which usually has a much lower boiling point than the solute, begins to boil and change into a gas. Thus the solvent **EVAPORATES** leaving behind the solute.

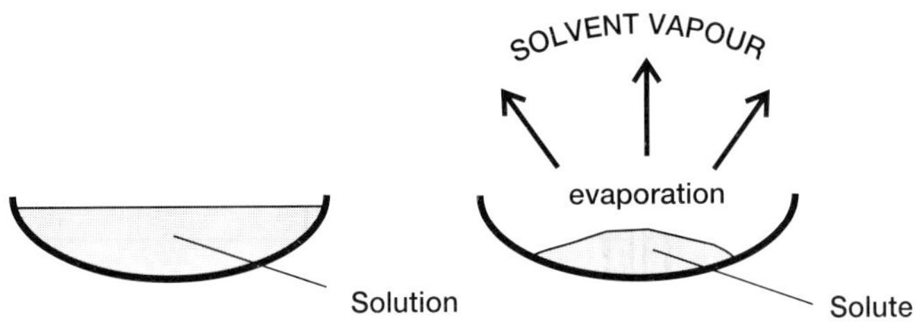


Fig. 2.4 *Evaporation*

Evaporation may be speeded up by heating the solution.

**NOTE**

> If your evaporating basin, watch glass, or test tube is completely clean when you have evaporated all the liquid from it, then what you had to start with was a pure liquid and not a solution.
>
> Thus evaporation can be used to test the purity of a liquid.

## Recovering the solvent from a solution

The process is in two parts:

1) **Evaporation** – to remove the solvent from the solution
2) **Condensation** – to change the solvent vapour into pure solvent

Let us consider a mixture of water and ethanol:

Ethanol has a boiling point of 78°C
Water has a boiling point of 100°C

In theory, if you heat a mixture of the two, then the ethanol would boil first and change to vapour at 78°C whilst the water stays as a liquid. The ethanol could then be collected first and the water second.

In fact, this does not happen. What actually happens is that when the ethanol/water mix is heated at 78°C, the vapour contains **mostly** ethanol together with some water vapour. If this vapour mix is condensed and re-distilled, then the mixture becomes more ethanol rich and so on. After many distillations, the ethanol becomes **almost** pure – in fact it is not possible to achieve 100% purity of this mix using distillation.

A way of achieving many 'mini-distillations' is by using a **FRACTIONATING COLUMN.** These come in different shapes, the simplest being a tube filled with glass beads.

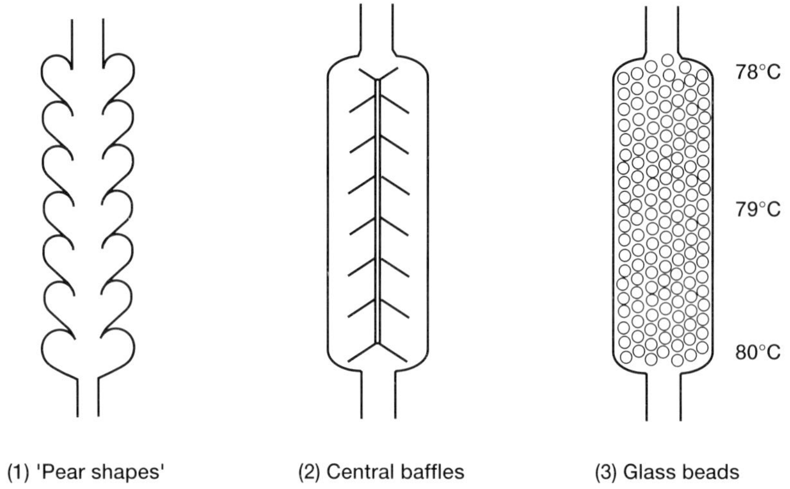


Fig. 2.7 *Some types of fractionating columns*

Looking at Fig. 2.7 (3), you will see that as the temperature of the column falls, the further you are from the distillation flask. So, in the case of the ethanol/water mix, as the mixed vapours rise up the column, the alcohol vapour travels upwards (as the temperature is greater than 78°C), whilst the water vapour condenses on the beads (as their temperature is less than 100°C), and falls back downwards. As the vapour mix rises up the column, the mix becomes more 'ethanol rich'.

Fractional distillation is an important part of the **oil refining** process. Oil as it comes out of the ground (known as **crude oil**) is a mixture of liquids which are all compounds of carbon. As the boiling points of each liquid are close to each other, batches of mixed liquids are collected at various points up the **fractionating tower.** These batches of liquids are called **fractions** and they are then re-distilled in other specialised fractionating towers to obtain the different materials, such as various oils, petrol and gases which we use in our everyday lives.

Air is a mixture of gases (see later) and these are also collected individually by fractional distillation. Air is cooled and compressed until all of the gases are in liquid form. The liquid mix is fractionally distilled and individual gases are collected as the liquid mix is heated, until the particular gases to be collected reach their boiling points (more of this later).

## Separating solids from mixtures

### SUSPENSIONS

You will remember that suspensions are mixtures of particles of INSOLUBLE SOLIDS 'suspended' in a liquid.

The most usual way of recovering the solids from a suspension is by **FILTRATION.**

| ONLY SOLUTIONS AND PURE LIQUIDS PASS THROUGH FILTER PAPER |
|---|

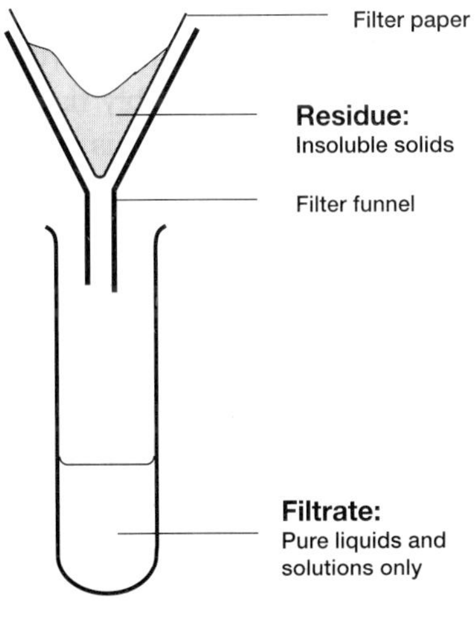


Fig. 2.8 *Filtration*

2) Police – to match up samples from a suspect and his clothing to substances found at the scene of a crime.
3) Biochemists – to separate proteins into their various parts in their constant search for drugs to fight disease and illness of all sorts.

*How to obtain separate samples from a chromatogram*
Remember that you are dealing with *very* small amounts of material, so your normal chemical apparatus will be too large and the substances would be 'lost' either in filter paper or spread out around the glassware so thinly that practical collection is made very difficult. This activity would be carried out using specialised 'micro' apparatus.

However, the principle of the separation would be the same and this is outlined below in Fig. 2.10.

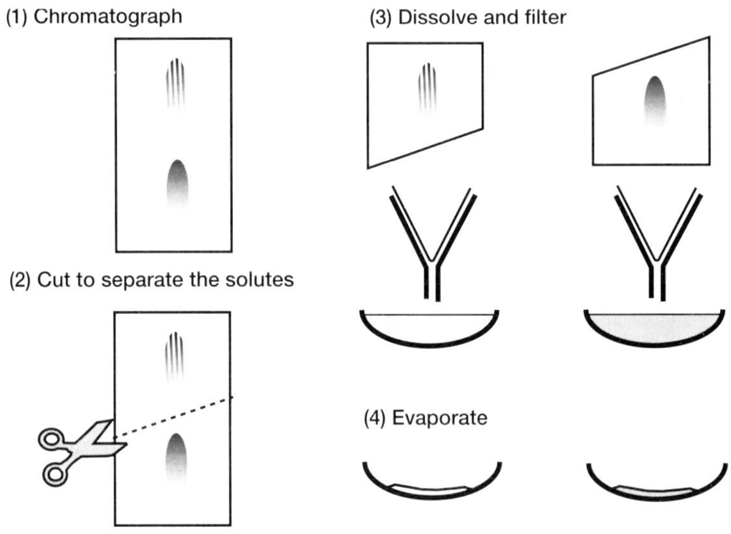


Fig. 2.10

## Separating two liquids which do not mix (are immiscible)

**REMEMBER**

EACH LIQUID HAS **ITS OWN** DENSITY.

So, as the liquids do not mix, the less dense liquid will float on top of the more dense one.

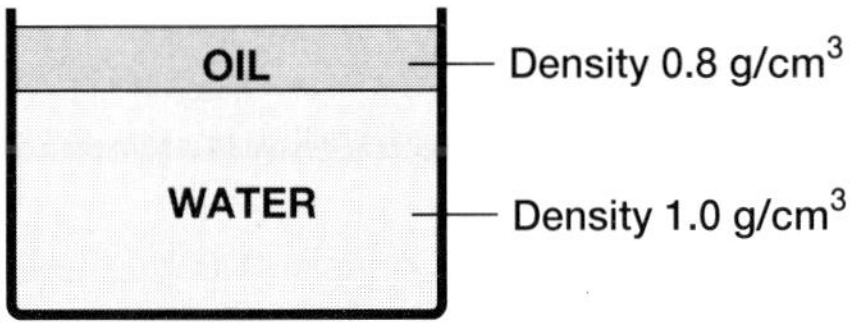


Fig. 2.11

A convenient way to separate these liquids is by using a **separating funnel**. See Fig. 2.12 below.

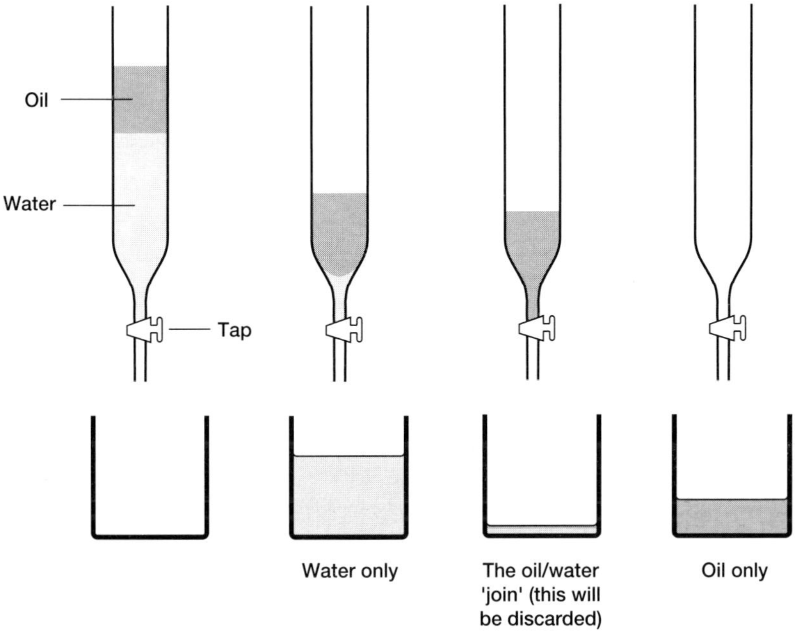


Fig. 2.12

# What do you know about ...

1. What do the following words mean?

   Solvent Solute Solution Soluble Insoluble

2. Draw a table with the headings ELEMENTS, COMPOUNDS, MIXTURES. Put the following substances in the correct columns:

   Air, carbon, carbon dioxide, distilled water, sea water, iron filings, crude oil, concentrated sulphuric acid, dilute sulphuric acid, magnesium, brass, sodium chloride

3 What is a 'saturated solution'? How can you tell that a solution is saturated?

4. 50 $cm^3$ each of tap water, sea water and distilled water were put into three evaporating basins **x**, **y** and **z**. The empty basins had been weighed **before** the addition of the samples. The liquids were then evaporated and the basins weighed again.

| BASIN | X | Y | Z |
|---|---|---|---|
| Mass of empty basin (g) | 66.3 | 66.1 | 66.2 |
| Mass of basin after evaporation of sample (g) | 68.4 | 66.1 | 66.4 |

   (a) Which type of water was evaporated in basins **x**, **y** and **z**?
   (b) Which of the three types of water does rain resemble? Explain your answer.

5. The label below is part of a label from a bottle of mineral water.

| **TYPICAL ANALYSIS mg/l** | |
|---|---|
| POTASSIUM | 4.6 |
| CALCIUM | 7.5 |
| SODIUM | 6.0 |
| MAGNESIUM | 1.8 |
| SILICA | 53.0 |
| BICARBONATE | 55.0 |

   (a) How do you know that this mineral water is not pure water?
   (b) How do calcium, magnesium and so on, find their way into the water?

# ... separating mixtures?

(c) (i) Name the method you would use to obtain pure water from mineral water?
(ii) How would you show that you had produced pure water?

(d) Some people prefer bottled water because they think that tap water is polluted.
(i) What is meant by the term 'polluted'?
(ii) What is the most important physical process used to change muddy river water into tap water?
(iii) What sort of impurity is **not** removed by the process?

6. Look at the following instruction labels on two different types of paint.

| **EMULSION PAINT** | **HIGH GLOSS PAINT** |
|---|---|
| USE: walls and ceilings<br>TO CLEAN BRUSHES:<br>use warm water | USE: all woodwork,<br>doors, windows etc.<br>TO CLEAN BRUSHES:<br>use white spirit |

(a) Whilst painting the walls, I spill some paint on my hands. What should I use to remove the paint?

(b) (i) Why will I use white spirit to clean out the brushes after painting the door?
(ii) Give a reason why I cannot use water instead.

(c) Show, using a labelled diagram to help you, how I might find out if the paint that I use on the walls contains just one pigment (colour) or is a mixture of pigments.

7. Hydrated copper sulphate is soluble in water. Sulphur is insoluble in water.

If you have a mixture of these two substances, describe how you would obtain some pure crystals of hydrated copper sulphate.

8. In England, salt is mined from the ground. However, the salt is mixed with quantities of rock as it comes to the surface. Describe how, on a laboratory scale, you can prepare a sample of pure salt from some rock salt. You need to describe **what** you would do and why you are doing it.

# What do you know about ...

9. In Mediterranean countries, salt can be obtained from sea water. Suggest how this might be done on a large scale, giving your reasons.

10. (a) Look at Figure 2.5 (page 26) and explain why you will not obtain a satisfactory distillation of ink using this apparatus.
    (b) Explain how you would overcome some of the problems encountered when using this apparatus.

11. You have a sample of sea water and you wish to obtain some pure water from it. Draw a labelled diagram to show how you would **best** achieve this.

12. Below is a chromatogram:

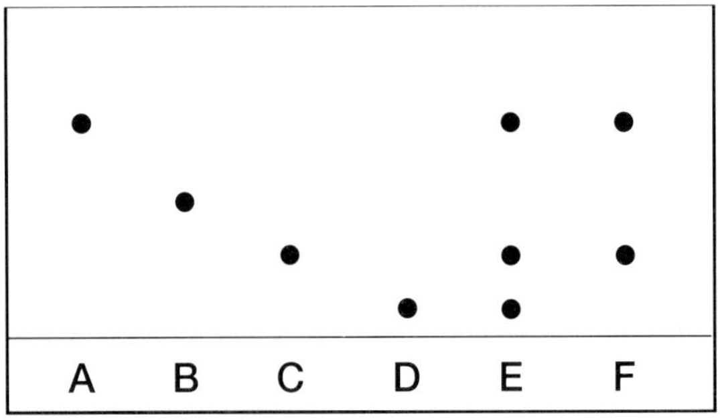


    (a) Which letters represent mixtures and which letters represent pure substances?
    (b) Which pure substance is not in any mixture?
    (c) Which pure substances are only in **one** mixture?

13. Some salad dressings are made of oil and vinegar – two liquids which are immiscible.

    (a) What does the word *immiscible* mean?
    (b) What happens if the dressing is allowed to stand unused for some time? Give a reason.
    (c) Show how you might separate the salad dressing into samples of pure oil and pure vinegar.

# ... separating mixtures?

# Further questions on ...

1. Below is a list of mixtures

   (i) sand and salt
   (ii) water and alcohol
   (iii) the colours in a universal (full range) indicator
   (iv) water and salt
   (v) iron filings and sand
   (vi) wood dust and salt

   (a) For each pair, name the process by which you would separate them.
   (b) Explain the physical property on which the separation method in each case depends.

2. Crude oil is fractionally distilled to produce a range of fractions which differ from each other.

   Explain, giving your reasons:

   (a) whether crude oil is a mixture or a compound;
   (b) why fractional distillation is used rather than simple distillation.

3 The diagrams and graphs on the following page refer to two separate distillations of a mixture of water and PROPANONE.

   (a) Why is there a faster temperature rise in thermometer A than in thermometer B?
   (b) When thermometer A reaches 100°C why does it remain constant for a while?
   (c) What happens at point R?
   (d) Why does thermometer C take much larger to reach 56°C than thermometer B?
   (e) Why does thermometer C remain steady, first at 56°C and then later at 100°C?
   (f) What is happening in the column of glass beads from point P to point Q?
   (g) Which apparatus would be more successful at separating propanone from water?

# ... separating mixtures

# Further questions on ...

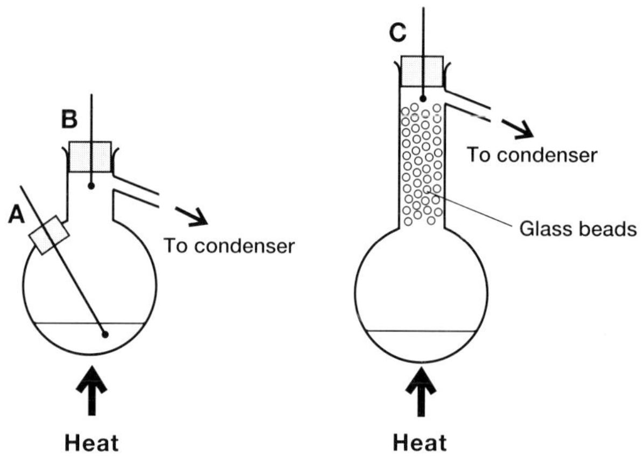


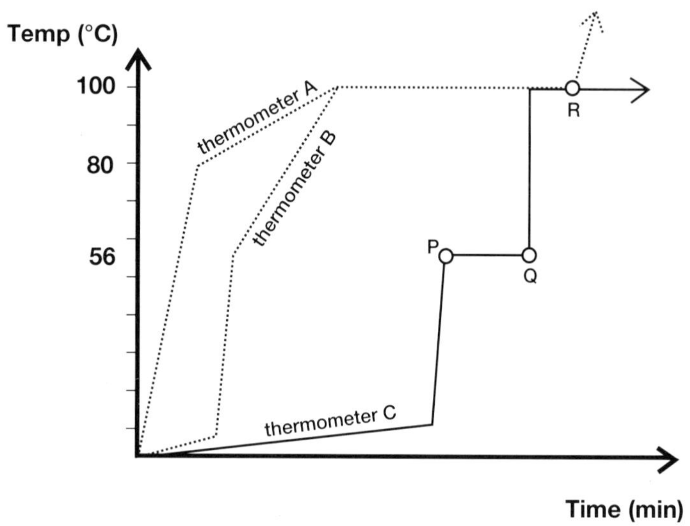


... separating mixtures

# Further questions on ...

4 The table below gives the solubility of lead nitrate in water at various temperatures.

| TEMPERATURE (°C) | 20 | 40 | 60 | 80 | 100 |
|---|---|---|---|---|---|
| **SOLUBILITY (g/100 g of water)** | **58** | **76** | **95** | **112** | **131** |

(a) Draw this information on a graph. Use the x-axis for the temperature.
(b) From your graph, determine the solubility of lead nitrate at (a) 10°C, (b) 50°C
(c) What mass of lead nitrate crystals would be formed if a saturated solution were cooled from 50°C to 10°C?
(d) Calculate the change in solubility of lead nitrate per degree Celsius over the temperature range 10°C to 100°C.
(e) A saturated solution of lead nitrate was left under pressure at 150°C. What mass of dissolved lead nitrate would it contain?
(f) Why must the solution be left under pressure?

---

**In answering question 5 below, you may NOT use stencils to do your drawings and NO writing other than to label parts of your diagram is allowed.**

---

5. For each of the mixtures below, draw a diagram of the apparatus you would use to separate the mixture into its parts.

(a) a solution of lead nitrate in water
(b) tar and sand into clean sand and tar
(c) black water-soluble felt-tip pen into the four colours making it up

6. Write an essay which is to be read by someone who knows no science. You must use simple words which explain the meaning of any scientific terms you use.

The title of the essay is:

'The principles used in separating impure substances'

## ... separating mixtures

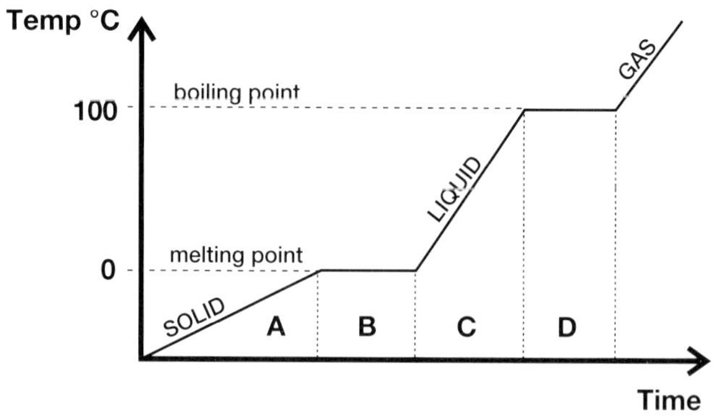


Fig. 3.1 *Melting of ice, boiling of water*

*During Time interval A*: Molecules are held in a framework known as a **lattice**. The lattice expands as the molecules move further apart.

*During time interval B*: Molecules break free of the lattice in **clumps**. The clumps start to move around each other as the solid becomes liquid. As it requires energy to break the lattice binding, there is **no increase** in **overall temperature until the melting is complete.**

*During time interval C*: The molecules and clumps of molecules move faster and further apart as the liquid becomes hotter.

*During time interval D*: Molecules break free from the surface of the liquid. Bubbles are seen in a boiling liquid because of particles and groups of particles, moving in opposite directions which leave empty spaces. (i.e. bubbles). As in changing state from solid→liquid, energy is required for particles to escape, so there is no increase in overall temperature until **all** the molecules have escaped i.e. boiling is complete.

Once the molecules have escaped from the liquid and the change of state from liquid to gas is complete, then the gas will become hotter if heating continues.

## Why are melting point and boiling point used to show that a substance is pure?

1. Pure substances will contain only **one** type of molecule.
2. The amount of energy needed to 'break' a solid lattice, **depends upon the type of molecule.** Each type of molecule will require a 'special' amount of energy to break the lattice and this is why each substance has its own boiling point and melting point.

3. Adding **another** type of molecule, as in an impurity, would change the arrangement of the lattice and thus the amount of energy needed to change the state of the substance.

| EACH PURE SUBSTANCE HAS **ITS OWN** BOILING POINT AND MELTING POINT |
|---|

## Increase in solubility

We have seen that an increase in the temperature of a substance causes an increase in the **activity** of the molecules of that substance.

So when solutions are warmed, the molecules of the **solvent** become more active and the dissolving process becomes 'speeded up'.

Thus solutes will **dissolve more quickly** when the solvent is **warmed.**

(Solutes will also dissolve more quickly when their **surface area** is increased by breaking them up into smaller lumps – generally until they are in powder form.)

## Chemical effects of heating

**REMEMBER**

A **chemical** change means that a **new substance** has been made. The substance(s) which you **start with** is/are called **REACTANT(S).**

The substance(s) which are **produced** as a result of a chemical reaction is/are called **PRODUCT(S)**.

Thus, chemical reactions can be written down as **WORD EQUATIONS** which give the names of the substances involved.

Sometimes, the **physical state** of the substances is shown using lower case letters (known as **state symbols)** in the word equation.

**FOR EXAMPLE**

| **state symbol** | |
|---|---|
| (s) | SOLID |
| (l) | LIQUID |
| (g) | GAS |
| (aq) | AQUEOUS (or 'in solution') |

**FORMING BONDS RELEASES ENERGY**

Let us look at some examples.

## HYDRATED COPPER SULPHATE

From your studies, you will know that when you heat hydrated copper sulphate, the blue crystals decompose into the white powder, anhydrous copper sulphate and water.

*In energy terms*: An **input** of energy is required to decompose hydrated copper sulphate.

Also, you will know that when water is added to **white anhydrous copper sulphate** powder:

1) the powder turns blue
2) the newly formed blue crystals become hot.

*In energy terms*: Forming hydrated copper sulphate from a combination of water and the anhydrous form, **releases** energy.

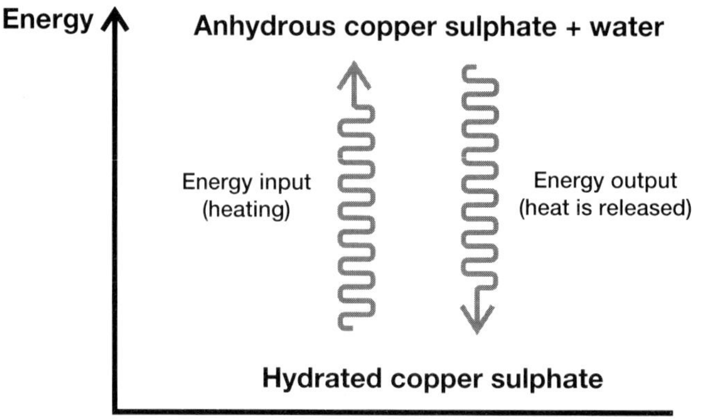


Fig. 3.3 *Anhydrous copper sulphate and water*

We can represent both of these reactions as a **single reversible reaction:**

$$\text{HYDRATED COPPER SULPHATE (s)} + \text{ENERGY (HEAT)} \rightleftharpoons \text{ANHYDROUS COPPER SULPHATE (s)} + \text{WATER (g)}$$

*Hydrated:* A substance which contains **water of crystallisation**, i.e. the water needed to make the crystal.

*Anhydrous:* A substance which contains no water, but is able to change into the crystalline line (hydrated) state if water is supplied. It is **not** enough merely to say 'a substance that contains no water'. Dry copper oxide contains no water but it is not anhydrous. If you add water to dry copper oxide it merely becomes wet.

**ENDOTHERMIC AND EXOTHERMIC REACTIONS**

If a chemical reaction requires an **INPUT** of energy. i.e. **takes *in* energy from the surroundings,** then the reaction is said to be **ENDOTHERMIC.**

If a chemical reaction produces an OUTPUT of energy, i.e. **releases energy *out to* the surroundings,** then the reaction is said to be **EXOTHERMIC.**

Some familiar exothermic reactions:

**TABLE 3.1**

| Reactants | Energy released to surroundings |
|---|---|
| Magnesium + oxygen | Heat, light |
| Methane (N. Sea gas) + oxygen | Heat, light, sound |
| Fuel (wood, coal) + oxygen | Heat, light, sound |
| Fuel (petrol) + oxygen | Heat, light, sound, movement |

## Heating and burning

We started the chapter by saying that heating and burning were different things. Now we are ready to look at why they are different.

Let us consider the exothermic reaction between magnesium and oxygen.

You will know that there are two stages to this reaction.

(i) *Stage I*: Magnesium ribbon is held in some tongs and **heated** over a Bunsen flame.
(ii) *Stage II*: A brilliant white flame is seen as the magnesium **burns** to form the white ash magnesium oxide.

Stage II is clearly exothermic. We have seen before that **forming** bonds, as in this case where bonds are formed between magnesium and oxygen, results in energy being released (see Fig. 3.4).

# What do you know about ...

10. Malachite contains copper carbonate which decomposes into copper oxide and carbon dioxide.

    (a) Make a list of the apparatus which you would need in order to find the density of carbon dioxide.
    (b) Draw a labelled diagram to show how you would find the density of carbon dioxide.
    (c) What measurements would you make and how would you use these to find the density of carbon dioxide.
    (d) Giving reasons, say what precautions you would take to make sure that your final result was as accurate as possible.

11. Explain, giving an example, what the terms 'exothermic reaction' and 'activation energy' mean.

12. A student heated some potassium manganate (VII) (potassium permanganate) in a test tube. The tube and its contents are weighed every minute. The weighings are given below.

| Time (min) | 0 | 1 | 2 | 3 | 4 | 5 |
|---|---|---|---|---|---|---|
| Mass of test tube + Potassium manganese (VII) (g) | 7.86 | 7.83 | 7.75 | 7.63 | 7.63 | 7.63 |

    (a) Plot a graph which shows how the mass changes with time.
    (b) How long does the reaction take to complete? Give a reason for your answer.
    (c) Explain, giving reasons, what sort of chemical reaction has taken place.

13. When blue crystals of hydrated copper sulphate are heated, the crystals change to a white powder (called anhydrous copper sulphate) and water vapour is given off. An experiment was carried out to find out how much water there is in crystals of hydrated copper sulphate. The results of the weighings are given below.

| | | |
|---|---|---|
| Mass of container | = | 18.30 g |
| Mass of container + hydrated copper sulphate | = | 23.30 g |
| Mass of container + anhydrous copper sulphate | = | 21.50 g |

    (a) Write a word equation which describes the thermal decomposition of hydrated copper sulphate.
    (b) Why do we call this reaction a 'thermal decomposition?'

## ... heating substances?

(c) (i) How much water is there in the crystals in this experiment? (show your working).
(ii) How much water would there be in 100 g of copper sulphate crystals?

(d) Express the amount of water which you have found in (iii) (a) and (b) as percentages.

(e) What do you notice about your answers in part (iv)? Use this figure to find out how much water there is in 1 kg of copper sulphate crystals.

(f) Draw the apparatus which you might use to collect the water given off.

14. Five sets of pupils set out to find how much carbon dioxide is produced when copper carbonate is heated.

They weighed samples of copper carbonate before and after heating. They also measured the volume of carbon dioxide being produced.

The results from their experiments are tabled below:

| Mass of tube and copper carbonate before heating (g) | Mass of tube and copper oxide after heating (g) | Volume of carbon dioxide collected ($cm^3$) |
|---|---|---|
| 5.53 | 5.50 | 16 |
| 4.86 | 4.80 | 33 |
| 6.10 | 5.91 | 49 |
| 6.85 | 6.73 | 75 |
| 7.00 | 6.85 | 82 |

(a) Produce a table which shows the loss of mass and carbon dioxide collected for each set of results.

(b) Draw a graph showing loss of mass (horizontal axis) against carbon dioxide produced.

(c) Which set of pupils' results is clearly wrong? Give at least two reasons why this particular set of results is 'out of line' with the rest of the class.

(d) (i) What would the pupils see happening in the test tube?
(ii) Write a word equation to describe the reaction which takes place in this experiment.

(e) From these figures, calculate a value for the density of carbon dioxide.

(f) Use the graphs to find out how much copper oxide would be remaining if you heated 2.34 g of copper carbonate which produced 10 $cm^3$ of carbon dioxide.

# Further questions on ...

6.

| SUBSTANCE | WHAT IS OBSERVED ON HEATING |
|---|---|
| Calcium carbonate | The white lumps do not seem to change at all |
| Copper | The pink shiny metal became hot and the Bunsen flame turns green. A black coating forms on the foil when it cools. |
| Magnesium | The silver metal burns with a very bright light. A white solid is left. |
| Magnesium oxide | The white powder does not seem to change at all. |
| Red ink | The ink boils. Eventually, all the liquid disappears leaving behind a red solid. |

(a) (i) Choose one substance from the list and explain why its mass would increase when heated in air.
(ii) Would this substance show the same or different mass change if it was heated in an unreactive gas such as argon instead of air? Explain why.

(b) (i) Write down, from the list above, the name of one element, one compound and one mixture.
(ii) Use the evidence in the table to explain the choices which you have made in part (i).

(c) Both calcium carbonate and magnesium oxide seem to produce similar results when heated. Is this true? If so, give a reason for your answer. If not, then say why it is not and give details of further experiments which need to be done to justify your answer.

7. Anhydrous copper sulphate is a white solid which is not affected by moderate heating. Copper carbonate is a green solid.

A mixture of anhydrous copper sulphate and copper carbonate was gently heated and the volume of gas which was given off was measured.

| Time (seconds) | 0 | 10 | 20 | 30 | 40 | 50 | 60 | 70 | 80 | 90 |
|---|---|---|---|---|---|---|---|---|---|---|
| Volume of gas ($cm^3$) | 0 | 3.5 | 9.0 | 16.2 | 26.9 | 36.6 | 43.7 | 47.6 | 50.0 | 50.0 |

## ... heating substances

(a) Plot a graph of these results.
(b) At what time was the gas coming off fastest? Give a reason.
(c) Which substance was decomposing as it was being heated?
(d) Will the solid mixture at the end of the experiment weigh more, the same or less than the original mixture? Explain your answer.
(e) What colour do you think the mixture will be at the end of the reaction after it has cooled down? Give a reason for your answer.

8. A group of pupils heated different masses of magnesium carbonate in identical test tubes until decomposition was complete. The cooled tubes and contents were reweighed and the decrease in mass calculated.

Their results are shown below:

| Mass of test tube + magnesium carbonate | Mass of test tube + residue |
|---|---|
| 18.8 g | 16.8 g |
| 21.0 g | 17.8 g |
| 24.6 g | 19.6 g |
| 27.0 g | 20.8 g |
| 30.0 g | 22.2 g |

(a) Copy out the table and add a further column headed 'Decrease in mass'. Calculate and record the various decreases in mass.
(b) Plot a graph of 'mass of test tube + magnesium carbonate', against 'decrease in mass' (horizontal axis).
(c) Use your graph to find the mass of the test tubes used. Explain how you obtained your answer.
(d) Calculate how much magnesium carbonate you would need to produce 20 g of residue.
(e) Write a word equation for the reaction which occurs during heating.

# Further questions on ...

9. Magnesium carbonate was put into a test tube and weighed. It had a mass of 14.34 g. It was heated and the volume of gas produced was recorded at various times. When the test tube had cooled down, it was weighed again – its mass was 14.14 g.
   The readings for the volume of gas produced at various times are given below.

| Time (seconds) | 0 | 30 | 60 | 90 | 120 | 150 | 180 | 210 | 240 |
|---|---|---|---|---|---|---|---|---|---|
| Volume of gas ($cm^3$) | 0 | 34 | 56 | 74 | 86 | 94 | 98 | 100 | 100 |

(a) Plot a graph to show how the volume of gas produced changes with time.
(b) How long does it take for half the magnesium carbonate to decompose?
(c) How might you increase the rate of decomposition? Make sure that you explain why the steps you would take, would be effective.
(d) Calculate the density of the gas in $g/cm^3$.

# ... heating substances

# INDICATORS AND THEIR USE IN CHEMISTRY

# 4

## What are indicators?

These are substances which change their appearance (e.g. colour), when they come into contact with particular substances.

In other words, they **indicate** that a particular substance is present.

## Limewater as the indicator to show that carbon dioxide is present

We have already seen that the gas carbon dioxide is given off when metal carbonates are heated.

We **prove** that the gas **is** carbon dioxide by testing with **limewater**.

Limewater is a clear solution of calcium hydroxide (also called lime) in water. Because the calcium hydroxide is in solution, i.e. has dissolved, we cannot see it.

When carbon dioxide is added to limewater, the gas changes the calcium hydroxide to calcium carbonate which is insoluble in water. As the calcium carbonate is being produced, small particles of it appear as a white **precipitate**, so we say that the limewater has gone 'cloudy' or 'milky' or, preferably, 'chalky' – because chalk is another name for calcium carbonate.

We can summarise the limewater test with the word equation:

| CARBON DIOXIDE (g) | + | LIMEWATER (aq) | ⟶ | CHALK (s) | + | WATER (l) |
|---|---|---|---|---|---|---|
| | | (Calcium hydroxide)<br>Soluble, so it is invisible | | (Calcium carbonate)<br>Insoluble, so it appears as a white precipitate | | |

**IF CLEAR LIMEWATER GOES CHALKY, THEN CARBON DIOXIDE IS PRESENT**

**THIS IS THE STANDARD TEST TO IDENTIFY CARBON DIOXIDE.**

## Two indicators which show that water is present

### ANHYDROUS COPPER SULPHATE

This changes from **white** to **blue** (and gives out heat) when it comes in contact with water.

| | | | | | | |
|---|---|---|---|---|---|---|
| ANHYDROUS COPPER SULPHATE (s) | + | WATER (l) | ⟶ | HYDRATED COPPER SULPHATE (s) | + | HEAT |
| (White powder) | | | | (Blue crystals) | | |

### ANHYDROUS COBALT CHLORIDE

This changes from **BLUE** to **PINK** (and gives out heat) when it comes in contact with water

| | | | | | | |
|---|---|---|---|---|---|---|
| ANHYDROUS COBALT CHLORIDE (s) | + | WATER (l) | ⟶ | HYDRATED COBALT CHLORIDE (s) | + | HEAT |
| (Blue powder) | | | | (Pink crystals) | | |

You may well have used **COBALT CHLORIDE PAPER** to test for water. In which case, you will know that the paper appears pink when it has just been torn out of a book of test papers.

This is because there is water vapour in the atmosphere which turns the paper pink. This means that you need to warm the paper gently to drive off this water and turn the paper **blue, before** you use it to test for water.

THESE TWO INDICATORS **ONLY** SHOW THAT WATER IS PRESENT.
THEY ARE **NOT** TESTS FOR PURE WATER

**TABLE 4.1 ADDING ANHYDROUS COPPER SULPHATE + ANHYDROUS COBALT CHLORIDE TO VARIOUS LIQUIDS**

| | Concentrated sulphuric acid | Dilute Sulphuric acid | Water | Salt solution | Propane |
|---|---|---|---|---|---|
| Adding anhydrous copper sulphate | stays white | turns blue | turns blue | turns blue | stays white |
| Adding anhydrous cobalt chloride | stays blue | turns pink | turns pink | turns pink | turns white |
| Conclusions from the experiments | no water present | **all** these **contain** water | | | no water present |

You will see that whilst **three** of the liquids **contain** water – only **one** of them (water), will boil at 100°C.

**SO:**

1. **ANHYDROUS COPPER SULPHATE + ANHYDROUS COBALT CHLORIDE WILL INDICATE THAT WATER IS PRESENT.**
2. **THE TEST FOR PURE WATER IS TO SEE IF IT BOILS AT 100°C.**

## Indicators to show that liquids are either acids or alkalis or neutral

Certain liquids (e.g. vinegar, lemon juice) have a sharp, sour taste and we can show these to be **acids** (from the latin word. *acidus – sour*). As one **never** eats or drinks in a laboratory, clearly taste is not a recommended way of testing liquids, so we use **indicators**.

In the laboratory we know that some liquids are acids, because the labels say so! You may well have come across **hydrochloric acid, sulphuric acid** and **nitric acid.** Again, we can show that these liquids are acids by using indicators.

### LITMUS – AN INDICATOR WHICH DETECTS ACIDS

Many indicators are coloured dyes which are extracted from plants. Litmus is extracted from a type of lichen and in water it is a purple solution which turns **red** when it comes in **contact with acids**.

**ALL ACIDS TURN LITMUS RED**

**PROBLEM!**

Litmus will turn red in **all acids**. This means that if you had an unlabelled bottle of white vinegar and another unlabelled bottle of sulphuric acid, adding litmus would not tell you which to put on your chips, for they would both turn litmus red. Clearly one acid is 'stronger' than the other.

There **is** a solution to the problem, but first we shall look at two other groups of liquids, **alkalis** and **neutral liquids**.

**ALKALIS**

There are liquids which are often referred to as the 'opposite' of acids. This is because they have the ability to 'cancel out' acids. This process of cancelling out acids is called **neutralisation** and we shall discuss this later on.

You can recognise an alkali in the laboratory because the labels with the chemical names on the bottle will end with the word – **hydroxide.** Thus limewater (calcium hydroxide) is an alkali.

Alkalis are often just as 'dangerous' as acids and care must be taken when handling these liquids. For example, sodium hydroxide is sometimes referred to as **caustic** soda. The word **caustic** means that it will chemically attack flesh causing injuries which are similar to burns.

However, some alkalis are much 'milder' and can be used to cure indigestion which is caused by too much acid in the stomach. The human stomach contains 'strong' acid which helps to break down the food we eat as part of the digestive system. However, sometimes we eat foods which are themselves acidic (e.g. too much vinegar on those chips) and this increases the amount of acid in our digestive system to a level which causes us discomfort – i.e. indigestion.

We can 'cure' this discomfort by taking medicines (either as tablets or in liquid form), which contain mild alkalis such as magnesium hydroxide mixed with calcium carbonate.*

## Litmus again! – a detector of alkalis

If a drop of the purple litmus solution is put into an alkali, then the litmus will turn **blue**.

This will be true for **all** alkalis, whether they be 'strong' i.e. sodium hydroxide, or 'weak', i.e. magnesium hydroxide.

| ALL ALKALIS TURN LITMUS BLUE |
|---|

## Neutral liquids

There are liquids which are **neither** acid or alkali. These are referred to as being **neutral**. **Water** is a good example of a neutral liquid.

* see later for how carbonates react with acids.

Some solutions are neutral and examples of these are **sugar solution** and **salt solution**.

If the purple litmus solution is mixed with a neutral solution **there is no change of colour.**

| NEUTRAL SOLUTIONS DO NOT CHANGE THE COLOUR OF LITMUS |
|---|

In your chemistry studies you may well have used litmus and so you know that it comes in three different ways:

1) as a purple solution of the dye in water
2) as test papers coloured red – red litmus
3) as test papers coloured blue – blue litmus

We can summarise the use of litmus as follows:

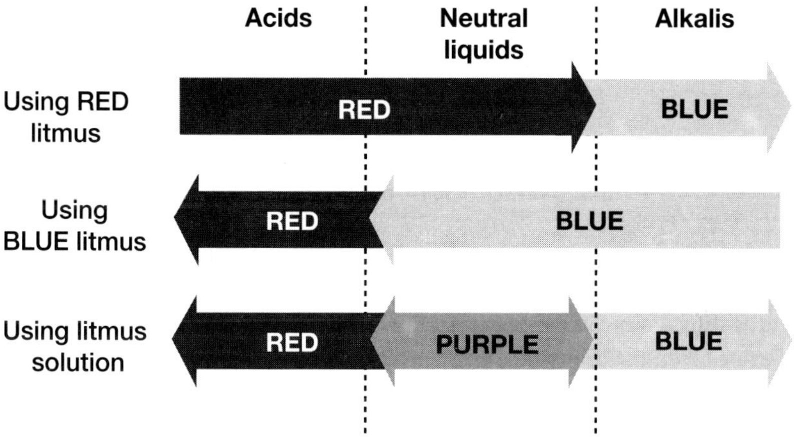


Fig. 4.2 *Testing with Litmus*

## Testing 'strength' of acids/alkalis – the pH scale

We have already seen that some acid/alkalis are 'strong', whereas others are 'weak'. Litmus will not give any idea of 'strength'. We measure 'strength' by means of the **pH Scale.**

The pH Scale runs from 0–14. There are very good mathematical reasons why these figures make up the scale, but we do not need to look into this further at this stage! Instead, we shall see how the scale works.

The middle of the Scale is 7 and **all solutions with a pH of 7 are neutral.**

| ALL SOLUTIONS WHICH HAVE A pH OF **LESS** THAN 7 ARE ACIDS |
|---|

| ALL SOLUTIONS WHICH HAVE A pH OF **MORE** THAN 7 ARE ALKALIS |
|---|

The weak acids/alkalis have pH values which are close to the mid-point (7), whereas the strong acids/alkalis have pH values close to the end points (0 and 14) of the scale. This is summarised in Fig. 4.3.

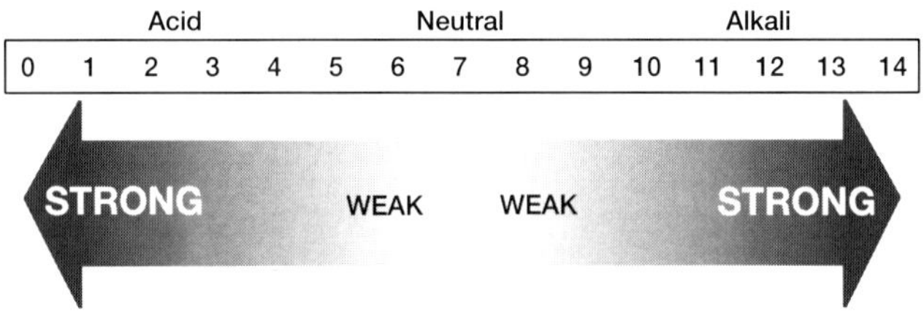


Fig. 4.3 *The pH Scale*

## HOW TO TEST THE PH OF A LIQUID

### 1. A pH meter

Quite simply – in theory – you dip a probe into the liquid and read the figure off the scale. In practice, reliable pH meters are fairly expensive and the procedures for using them can be quite complicated, so they are not often used at this stage.

### 2. Use of an indicator

There are indicators which are different colours at particular pH values. These indicators came either in liquid form or on test papers which have been soaked in the indicator.

An indicator which is able to change colours – each one representing a particular pH value – is known as a **UNIVERSAL INDICATOR**. There are many types of universal indicator, depending on the manufacturer, and so it is wise to check on the colour chart which is usually supplied with the indicator when determining the pH of solutions. Some types of universal indicator are called **FULL RANGE INDICATOR**, because a wide range of pH values can be detected when using it.

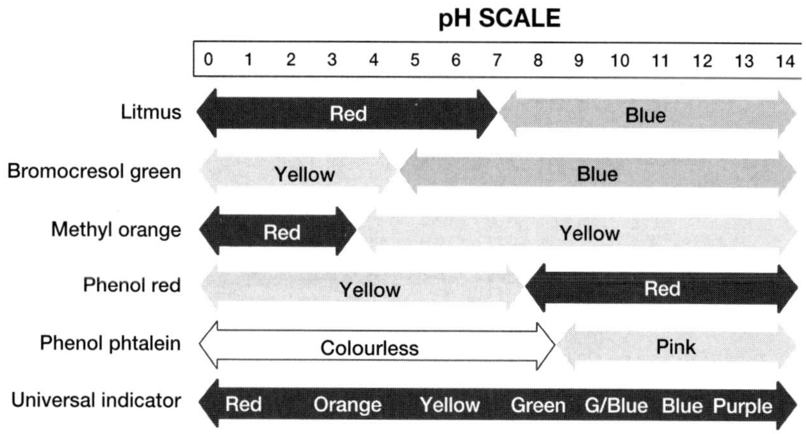


Fig. 4.4 *Indicators and pH value*

## What is a universal indicator made of?

You will remember that most indicators are made from plant extracts. Just as there is a great variety of plants, so there are hundreds of different indicators, each one changing colour for a particular pH value. Some examples are shown in Fig. 4.4.

You can probably gather from this figure that a universal indicator is a **mixture** of **dyes** which are usually dissolved in ethanol.

## Neutralisation

We have seen that an acid can be 'cancelled out' by adding an alkali to it and we have seen how this can be used to cure indigestion. This process of curing or cancelling out is really a chemical reaction called **neutralisation**.

Look at the experiment in Fig. 4.5, on the following page.

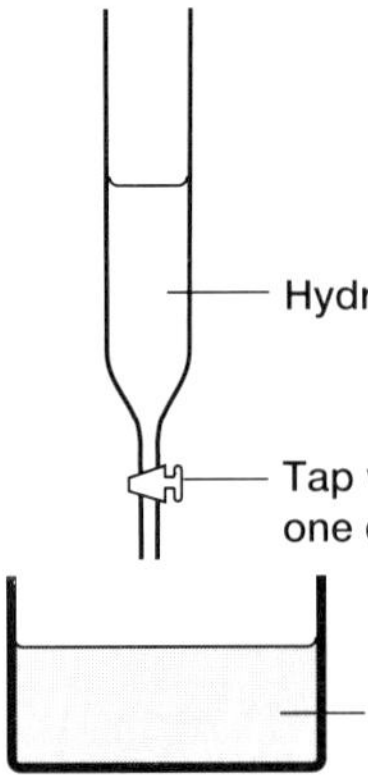


Fig. 4.5 *Adding acid drop by drop*

i) At first the solution stays purple
ii) The colour changes to blue
iii) The colour changes to **green** – **STOP THE ACID**

Now in the beaker, you have a **NEUTRAL SOLUTION**. If you feel the beaker, you will find that the beaker has warmed a little.

This shows that **energy** has been released by the reaction between the acid and the alkali and is further evidence that neutralisation is a **chemical reaction.**

If you take half of the now-green solution and evaporate it, you will find that small particles of a white appear. This white solid is **calcium chloride.**

The sample of calcium chloride has been made as one of the products of a neutralisation reaction, and it is called **a salt**. The other product of the reaction is the neutral liquid (hence **green** colour of indicator), **water.**

**THUS**

i) an acid – hydrochloric acid (also known as hydrogen chloride)
ii) an alkali – limewater (calcium hydroxide)

have reacted to form:

**A SALT** (calcium chloride) and **WATER** (the neutral liquid).

We can show this by the word equation:

HYDROGEN CHLORINE (*Hydrochloric acid*) + CALCIUM HYDROXIDE ⟶ CALCIUM CHLORIDE (*Limewater*) + WATER

**ACID** + **ALKALI** $\xrightarrow{\text{NEUTRALISATION}}$ **SALT** + **WATER**

Now back to the experiment in Fig. 4.5 with the remainder of the green liquid in the beaker.

(i) **Add *one* drop of acid** – The colour changes from green to yellow
(ii) **Add *one* more drop of acid** – The colour changes from yellow to orange.
(iii) **Adding acid drop by drop** – It takes a few drops of acid to change the colour from orange to red.

Once the colour is red, then there is no further change of colour – even if all the acid is dropped into the beaker.

You will notice that around neutral, a **very small** amount of acid caused a **large change** in pH. This can be shown by a graph:

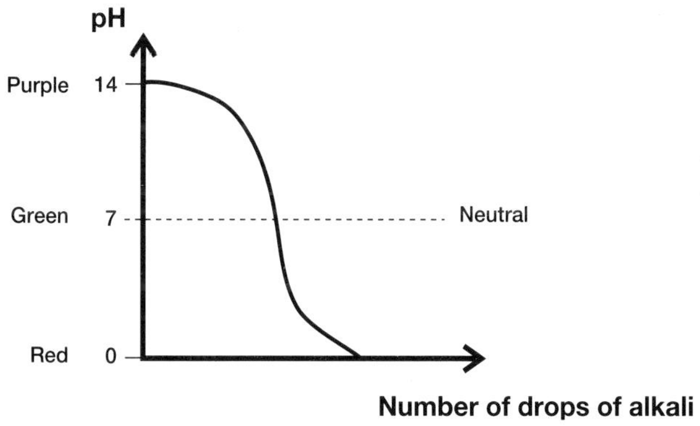


This experiment shows that:

**ADDING ACID TO AN ALKALI LOWERS THE pH OF THE SOLUTION**

If you repeated the experiment as shown in Fig. 4.5 but this time you put the alkali in the tap funnel and the acid in the beaker, you would start with a red solution in the beaker and go through the various colours until the solution in the beaker was purple.

This will:

(a) dilute the acid as it dissolves in the water;
(b) physically 'knock' the diluted acid off your hand and away down the sink.

**NEVER ADD AN ALKALI TO AN ACID SPILL – YOU DO NOT WANT A NEUTRALISATION REACTION TAKING PLACE ON YOUR HAND**

## More about salts

### ACIDS

These are all compounds of hydrogen and it is the **acid** which gives a salt its **second name**.

**FOR EXAMPLE**

| Name of acid | Chemical name of acid | Second name of salts produced |
|---|---|---|
| Hydrochloric acid | Hydrogen chloride | — chlorides |
| Sulphuric acid | Hydrogen sulphate | — sulphates |
| Nitric acid | Hydrogen nitrate | — nitrates |
| Carbonic acid | Hydrogen carbonate | — carbonates |

### ALKALIS

**ALL METAL OXIDES ARE CALLED BASES**

**Some** metal oxides dissolve in water to form **hydroxides.**

**FOR EXAMPLE**

| Base | Soluble in water | Alkali formed | 'Common name' |
|---|---|---|---|
| Calcium oxide | Yes | Calcium hydroxide | Lime |
| Copper oxide | No | – | – |
| Iron oxide | No | – | – |
| Potassium oxide | Yes | Potassium hydroxide | Caustic potash |
| Sodium oxide | Yes | Sodium hydroxide | Caustic soda |
| Zinc oxide | No | – | – |

When a **base**, or alkali, neutralises an acid, the **first name** of the **base** becomes the **first name** of the **salt.**

**SALTS**

You now know that:

| ACID + ALKALI ⟶ SALT + WATER |
|---|

Also:

| ACID + BASE ⟶ SALT + WATER |
|---|

**FOR EXAMPLE**

| Base or Alkali | Acid | Salt Formed |
|---|---|---|
| Copper oxide | Sulphuric acid | Copper sulphate |
| Copper oxide | Hydrochloric acid | Copper chloride |
| Copper oxide | Nitric acid | Copper nitrate |
| Sodium hydroxide | Hydrochloric acid | Sodium chloride* |

*Sodium chloride is often called **salt**. Now you know that it is one of **many** salts than can be produced as a result of a neutralising reaction. In order to avoid confusion, chemists often refer to sodium chloride as **common salt** or **table salt.**

## Carbonates can make salts

Any carbonate will fizz and release carbon dioxide when added to a dilute acid. In fact this is the test done to show that a substance **is** a carbonate.

What in fact is happening is that the carbonate is neutralising the acid.

You can show this by adding some calcium carbonate (small marble chips are convenient) to dilute hydrochloric acid which has some universal indicator in it. The solution is red to start with and fizzing can be seen as carbon chloride is given off. After a long time (a day or two) the liquid will have turned green as the reaction has finished and the acid has 'changed' to water.

If you pour the green solution into an evaporating basin and evaporate the water, then you will see the solid calcium chloride appear.

# What do you know about ...

5. If you have some blue litmus paper, what colour will it go when put into the following liquids:

   hydrochloric acid, limewater, sodium hydroxide, sugar solution, water

6. Which of the following liquids will turn red litmus blue?

   hydrochloric acid, limewater, sodium hydrochloride, sugar solution, water

7. Without looking up, draw a pH scale running from 0–14. On your diagram label:

   1) the neutral point
   2) acids – showing strongest and weakest
   3) alkali – showing strongest and weakest

8. Fred tested the following liquids with a universal indicator: vinegar, sodium hydroxide, sulphuric acid and distilled water.

   He wrote down the pH values which he obtained, but forgot to put in the names of the substances! Complete the table for him.

| Solution tested | pH value |
|---|---|
| | 1 |
| | 5 |
| | 7 |
| | 14 |

9. Using the apparatus as shown in Fig. 4.5 (page 68), 25 $cm^3$ of alkali was put into the beaker. Three different amounts of acid were dripped into the beaker. The pH value of the liquid in the beaker was measured after each drop and the following values were obtained:

   1 7 13

   The following amounts of acid were used:

   10 $cm^3$ 15 $cm^3$ 20 $cm^3$

   (a) Using the following headings, tabulate these results:

   amount of acid, amount of alkali, pH

## ... indicators, acids and alkalis?

(b) If you had put a piece of litmus paper in the beaker before you added the acid what colour would it have been?
(c) If you had added universal indicator to the beaker, say what colour it would have been for each of the pH readings which you took.
(d) Which was stronger – the acid or the alkali? Give a reason for your answer.

10. Samples of rainwater and distilled water were tested and found to have different pH values.

(a) How might they have measured the pH values of the samples?
(b) What values would be expected for
(i) distilled water; (ii) rainwater?
(c) What substance is likely to be present in rainwater to give it its different pH value?

11. A solution of bicarbonate of soda (sodium hydrogen carbonate) was added to some vinegar. The solution was added 2cm$^3$ at a time and after each addition, the mixture was stirred and the pH was measured. The results are shown below.

| Vol. of sodium hydrogen carbonate | 0 | 2 | 4 | 6 | 8 | 10 |
|---|---|---|---|---|---|---|
| pH of mixture | 3 | 5 | 7 | 8 | 8 | |

i) Draw a graph of these results (pH on vertical axis).
ii) How would you measure the pH?
iii) What was the pH of vinegar at the start of the experiment?
iv) How much sodium hydrogen carbonate was needed to produce a neutral solution?
v) Add an extra point to the graph to show the result of adding 10 cm$^3$ of sodium hydrogen carbonate.
vi) From this experiment can you deduce what the pH of sodium hydrogen carbonate is?

12. Rhubarb juice is acidic which makes it very sour. Sugar is often added to cooked rhubarb to stop it tasting sour.

Does the sugar actually react with the acid in the rhubarb, or does it just hide the taste?

Describe an experiment which you would do to find the answer to this question.

... indicators, acids and alkalis?

# What do you know about ...

13.

| | | | | | | | | | | | | | |
|---|---|---|---|---|---|---|---|---|---|---|---|---|---|
| 1 | 2 | 3 | 4 | 5 | 6 | 7 | 8 | 9 | 10 | 11 | 12 | (13) | 14 |

The dial of a pH meter is shown above. A light beam swings to and fro and shows the pH value of the liquid being measured.

In an experiment, the following liquids were tested: pure water, sodium hydroxide, vinegar and hydrochloric acid.

(a) Is the liquid being tested on acid or an alkali?
(b) Which of the four liquids named above could this be?
(c) Which of the four liquids is most likely to have a pH of **1**?
(d) Which liquid would you add to hydrochloric acid to neutralise it?

14. The naturally occurring substances that are present in flower petals often make good indicators for acid/alkali reactions.

(a) How would you obtain a useable indicator solution from red rose petals?
(b) Describe fully what you would do to compare it with a commercial universal indicator solution.

15. (a) What is the pH of pure water?
(b) What would you add to water to obtain solutions having pH values of:
(i) 12–13
(ii) 5–6
(iii) 1–2
(c) Explain one example of an everyday process where neutralisation takes place.

16. Complete the following:

i) ACID + BASE → ____ + WATER.
ii) ACID + ALKALI → ____ + WATER.
iii) ACID + CARBONATE → _______ + WATER + _______.
iv) ACID + METAL → ____ + HYDROGEN.
v) Adding acid to an alkali _____ the pH of the mixture.
vi) Neutralising an acid by a base _____ the pH of the mixture.
vii) Adding water to an acid _____ the pH of the acid.

## ... indicators, acids and alkalis?

# What do you know about ...

17. Some bases were added to water. The mixture was then filtered and a universal indicator was added to the filtrate. The results are shown below.

a) Copy and complete the table.

| Base | pH of filtrate | Is base soluble in water? | Name of alkali formed |
|---|---|---|---|
| Calcium oxide | alkaline | | |
| Copper oxide | neutral | | |
| Iron oxide | neutral | | |
| Magnesium oxide | alkaline | | |
| Sodium oxide | alkaline | | |
| Zinc oxide | neutral | | |

b) Write the names of the salts formed and the word equations for the reactions when each of the above bases neutralises sulphuric acid (hydrogen sulphate).

## ... indicators, acids and alkalis?

# Further questions on ...

1. Thermal blue is red in hydrochloric acid, but yellow in vinegar. Bromophenol blue is yellow when thymol blue is yellow, but blue in pure water.
Methyl red is red when bromophenol blue is yellow, but yellow in alkaline solution.
Phenolphthalein is colourless when bromophenol blue is blue, but red when methyl red is yellow.

   (a) Make a chart of these different indicator colour changes.
   (b) What colour would thermal blue be in pure water?
   (c) What colour is methyl red, when phenolphthalein is red?
   (d) What colour is bromophenol blue, when thermal red is blue?
   (e) What colour is phenolphthalein, when thermal blue is yellow?
   (f) What colour is phenolphthalein, when bromophenol blue is yellow?
   (g) What colour is methyl red in pure water?

2. Solution X has a pH of 1. Into a beaker is poured 10 $cm^3$ of solution X. Solution Y is now added to solution X a little at a time and the pH is checked after each addition. The results are given below:

| Vol. of Y added ($cm^3$) | 0 | 10 | 15 | 18 | 20 | 22 | 25 | 30 | 40 |
|---|---|---|---|---|---|---|---|---|---|
| pH | 1 | 1 | 1.2 | 1.8 | 7 | 12.6 | 13.4 | 14 | 14 |

   (a) Plot a graph of these values with pH on the vertical axis.
   (b) What type of chemical is:
      i) solution X
      ii) solution Y
   (c) Select from the list below a substance which could be X, and one which could be Y. Make sure you say which is which. Carbon, cobalt chloride, hyrdrochloric acid, magnesium, potassium permanganate, sodium hydroxide, sodium chloride, sugar, water.
   (d) What substances might be present in the beaker when 20 $cm^3$ of Y had been added?
   (e) Are solutions X and Y the same concentration? If not, explain which is the more concentrated, X or Y.
   (f) Explain why the final pH is 14 and why it has ceased to change.

# ... indicators, acids and alkalis

# Further questions on ...

3. A sample of soil was shaken with water and filtered to remove any undissolved solids. The pH of the filtrate obtained was found to be 6 and contained nitrates.

   (a) (i) Was the soil acidic or alkaline?

   (ii) Name a substance that might have been used to measure the pH of the soil.

   (iii) What could be added to the soil to make it neutral? Write a word equation for the reaction that would take place.

   The residue from the filtration was carefully dried and on treating with dilute hydrochloric acid, no reaction was observed.

   (b) What rock could **not** have been present in the soil? Explain your answer.

4. Suggest a chemical explanation for the fact that beetroot, when left in vinegar, gains a deep red solution, but that the colour changes when sodium hydrochloride is added.

5. A box contained some test tubes, a piece of glass tubing and five bottles of solutions. Only one of the bottles, sulphuric acid, had its label attached. In the bottom of the box were four labels: copper sulphate, sodium carbonate, hydrochloric acid and limewater.

   A student set out to identify the four liquids. As a start, he called the unlabelled bottles W, X, Y and Z.

   After a little while, he took bottle X and stuck on the label.

   Using the glass tube, he blew into samples of the other three liquids and was able to label bottle W.

   He added sulphuric acid to samples of each of the other two liquids and was able to correctly label Y and Z.

   What were the liquids W, X, Y and Z? For each liquid, say how you arrive at your conclusion, using the evidence provided by the student.

6. i) Explain the meaning of the words 'neutral' and 'salt'.

   ii) Must all salts be neutral?

   iii) Can salt on its own be neutral? If so how would you test to see if it was?

## ... indicators, acids and alkalis

# Further questions on ...

7. Tap water in certain areas contains dissolved substances which makes the water 'hard'. When hard water is boiled, calcium carbonate is slowly produced, forming solid deposits which are known as limescale.

   Three separate samples of limescale were tested and the results are given below:

| Test | Result |
|---|---|
| 1. Limescale + dilute hydrochloric acid | Bubbles formed as gas was given off |
| 2. Limescale was heated, cooled and dilute hydrochloric acid added to cold residue | No apparent reaction |
| 3. Limescale was heated, cooled, and added to distilled water. A few drops of universal indicator was added. | An alkaline solution was obtained. |

(a) Account for observations 1, 2 and 3, giving your reasons.

**It is possible to buy 'descaling liquid' which removes limescale by reacting with it to form carbon dioxide and a soluble calcium compound.**

(b) What can you say about the pH of 'descaling liquid'? Give a reason.

(c) Why is it a good idea that the calcium compound, formed by the action of 'descaling liquid' on limescale, is soluble?

(d) Write a word equation to show the reaction between 'descaling liquid' and limescale.

(e) Some limescale was added to a quantity of 'descaling liquid' which had been placed in a beaker on top of a balance. It was found that the total mass of the mixture decreased steadily for 3 hours and then became constant. Comment on the mass change.

# ... indicators, acids and alkalis

# Further questions on ...

7. A green powder (A), on being added to dilute sulphuric acid, fizzed and a gas (B), was given off, leaving a blue solution (C). When the gas was passed through liquid (D), the liquid changed from clear to chalky.

   Another example of (A) was heated. A gas was given off which turned liquid (D) chalky and a black solid (E) remained in the test tube.

   When diluted sulphuric acid was added to (E) and warmed, the black solid disappeared and a blue solution (C), was formed.

   It is known that (A) was a compound of copper.

   (a) Identify the substances A, B, C, D and E, giving your reasons.
   (b) Write word equations for the reactions:
       (i) A and dilute sulphuric acid.
       (ii) heating of A
       (iii) E and dilute sulphuric acid
   (c) Giving a reason, say which of A, B, C, D, E is:
       (i) a base
       (ii) a salt

8. Some marble chips (calcium carbonate) were put into a beaker of dilute hydrochloric acid which had been placed on the pan of a weighing machine. Mass readings were taken every minute and the results are given below:

| Time (min) | 0 | 1 | 2 | 3 | 4 | 5 | 6 | 7 | 8 | 9 |
|---|---|---|---|---|---|---|---|---|---|---|
| Mass (g) | 9.97 | 9.84 | 9.76 | 9.71 | 9.67 | 9.63 | 9.61 | 9.60 | 9.59 | 9.59 |

   (a) Plot a graph of mass against time (horizontal aims).
   (b) Calculate the change in mass during the following time spans:
       (i) 0.5–1.5 min
       (ii) 4.5–5.5 min
   (c) What do these results tell you about the value of this reaction?
   (d) (i) What is the change in mass during the ninth minute?
       (ii) What is the significance of your answer to (d)(i)?
   (e) Draw targets to the graph at one and four minutes and hence calculate the rates of reactions, i.e mass loss per minute, at these times.

## ... indicators, acids and alkalis

# Further questions on ...

9. When diluted hydrochloric acid is added to some zinc granules, hydrogen gas is produced as one of the products of the reaction.

   (a) If hydrochloric acid can also be called hydrogen chloride, what is the other product of the reaction? Write a word equation for the reaction.

   In an experiment to investigate the rate of this reaction, a gas syringe was used to collect the hydrogen and volume of measurements were made every half minute. The results are given below:

| Time (min) Vol of gas | 0.5 | 1 | 1.5 | 2 | 2.5 | 3 | 3.5 | 4 | 4.5 | 5 |
|---|---|---|---|---|---|---|---|---|---|---|
| Evolved ($cm^3$) | 22 | 34 | 44 | 50 | 55 | 58 | 59 | 60 | 60 | 60 |

   (b) Plot a graph to show how the volume of gas evolved changes with time.
   (c) What is the maximum volume of gas obtained in the experiment?
   (d) How long does it take to produce *half* the volume of gas?
   (e) Estimate the average rate ($cm^3$/min) at which the gas is produced between 2 mins and 2.5 min.
   (f) At what time is the reaction going fastest? Give a reason for your answer.
   (g) If you doubled the concentration of acid, describe how the graph would differ.

10. A beaker contains 20 $cm^3$ of dilute hydrochloric acid whose pH is 0.

    50 $cm^3$ of an alkali, sodium hydroxide, are steadily added to the beaker.

    The temperature and pH of the liquid in the beaker are measured regularly as the alkali is added. The readings are given below:

| Vol of alkali added ($cm^3$) | 0 | 10 | 20 | 30 | 40 | 50 |
|---|---|---|---|---|---|---|
| Temperature (°C) | 15 | 18 | 21 | 20 | 19 | 18 |
| pH | 0 | 0.5 | 1.0 | 13.0 | 13.5 | 13.5 |

## ... indicators, acids and alkalis

# Further questions on ...

(a) On the *top half* of a sheet of graph paper, plot the temperature of the mixture against the volume of alkali added (horizontal axis).
(b) On the *bottom half* of the same sheet of graph paper, plot the pH of the mixture against the volume of alkali added (horizontal axis).
(c) Is the final solution acidic, alkaline or neutral? Give a reason.
(d) Which is more concentrated, the acid or alkali? Give a reason.
(e) How much acid must be added to neutralise the 20 $cm^3$ of alkali?
(f) Write a word equation to describe the neutralising reaction taking place in this experiment.
(g) Why does the temperature rise and fall as the alkali is added?
(h) Why does the pH change in the way shown on the graph?

## ... indicators, acids and alkalis

## Gases can be impure substances: i.e. a mixture

**– a mixture**

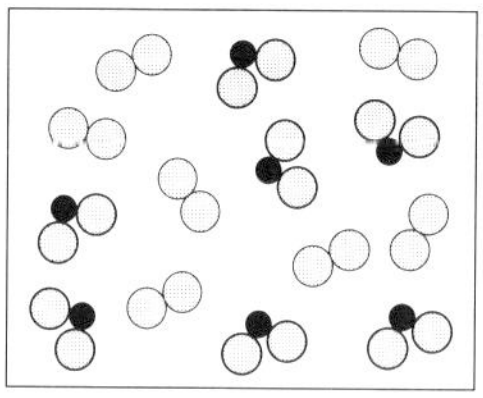

A mixture of
an element (oxygen) and
a compound (carbon dioxide)

Fig. 5.2

## Not all gases are invisible

**FOR EXAMPLE.**

| NAME OF GAS | COLOUR OF GAS |
|---|---|
| CARBON DIOXIDE | COLOURLESS |
| CHLORINE | GREEN |
| HYDROGEN | COLOURLESS |
| OXYGEN | COLOURLESS |
| NITROGEN | COLOURLESS |
| NITROGEN DIOXIDE | BROWN |

## Take care when measuring the volume of a gas!

### TEMPERATURE IS IMPORTANT

As a temperature increases, so does the kinetic energy of the gas particles. This means that they hit the sides of the container harder, or with greater force.

The combined forces of the molecules hitting the sides of a container results in what we call **GAS PRESSURE**

Thus, if the volume of the container remains the same:

**INCREASING THE TEMPERATURE OF A GAS INCREASES ITS PRESSURE**

The faster molecules of hot gas hit the sides of the container with greater force than the slower molecules of cool gas.

The gas syringes which you may have used in the laboratory are able to *change volume.*

Think of what is happening to the gas molecules as they are heated.

Their kinetic energy increases and so they hit the sides of the container with **greater force**. If the side of the container is able to move – as in the plunger of a gas syringe – then the **volume** of the container will **increase.**

As the volume increases, so the pressure remains the same.

So, when the pressure of a gas remains constant:

**INCREASING THE TEMPERATURE OF A GAS INCREASES ITS VOLUME**

**REMEMBER**

We are dealing with a **sealed container** which means that no molecules can get in or out – so the **mass remains** the **same**.

The relationship between the temperature and pressure of a gas was worked out by the Frenchman, **Jacques Charles.**

It was he who discovered that the volume of a gas increased in direct proportion to the rise in temperature – provided that the pressure of the gas remained the same.

Or, mathematically:

$$\text{Volume} \propto \text{Temperature}$$

$$\text{Therefore: } \frac{V}{T} \text{ is constant}$$

This is known as **Charles' Law** and you will use it to work out what volume a gas will occupy when the temperature rises, say from $T_{START}$ to $T_{END}$ – or more usually, $T_1$ to $T_2$.

## KEEPING THE TEMPERATURE THE SAME, BUT CHANGING THE VOLUME OF A GAS

Accurate work on this was first carried out by **James Boyle.**

You will know from your physics that if you **increase** the **volume** of a gas, you **decrease** its **pressure.**

**FOR EXAMPLE**

| **Experiment** | **1** | **2** |
|---|---|---|
| Volume ($cm^3$) | 4 | 10 |
| Pressure ($N/cm^2$) | 25 | 10 |

**NOTE**

$$P_1\ (25) \times V_1\ (4) = 100$$

and $$P_2\ (10) \times V_2\ (10) = 100$$

**THAT IS**

$$P_1 \times V_1 = P_2 \times V_2$$

This is known as **BOYLE'S LAW.**

Boyle's Law and Charles' Law are part of what are known as the **GAS LAWS.** You will meet these later.

## Gases in particular

### THE AIR

Up until about 250 years ago, air was considered to be a single substance. Indeed, the Greek philosopher Aristotle classified air as one of the four elements (air, water, fire, earth).

Until chemical tests are done, there is no reason why we should consider air as a mixture. After all, we cannot see it, or any of the component parts of the mixture.

### CHEMICAL EVIDENCE TO SHOW THAT AIR IS A MIXTURE

**Presence of water vapour**

Leave an open disk of **anhydrous copper sulphate** on a side bench. After a few hours, the white anhydrous copper sulphate will turn blue.

This shows that **WATER VAPOUR** is one of the gases of the air.

**Presence of carbon dioxide**

Leave an open disk of **limewater** on a side bench.

After a few hours a chalky crust will appear on the surface of the limewater.

This shows that **CARBON DIOXIDE** is one of the gases of the air.

**Presence of oxygen and nitrogen**
Look at the experiments below and on the following page.

Experiment 1 shows that air is **a mixture of at least two gases:**

$^1/_5$ used up in burning – oxygen
$^4/_5$ will not support combustion – nitrogen

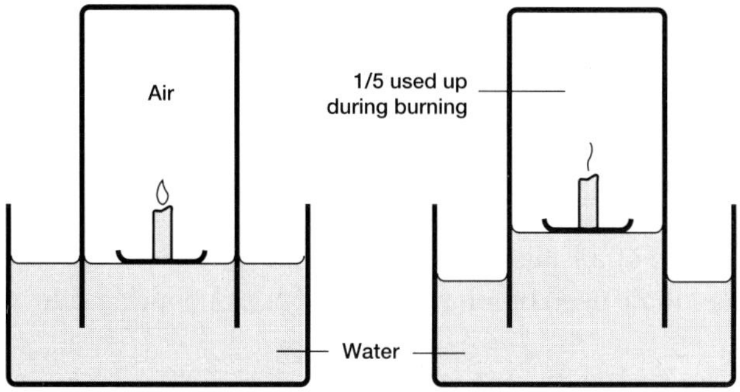


Fig. 5.3 *Experiment 1: burning*

**NOTE**

This experiment will only give us rough **approximations** of the volumes of oxygen and nitrogen in the air. We shall see later that other products are formed when a candle burns in air.

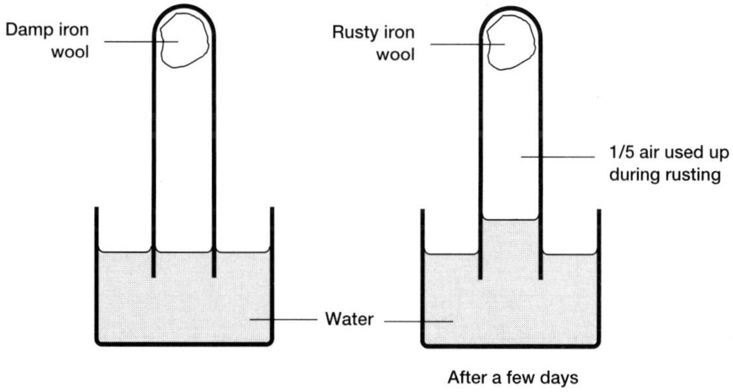


Fig. 5.4 *Experiment 2: rusting*

Again, $^1/_5$ of the air is used up as the iron rusts (oxygen).

The remaining gas puts out a lighted splint (nitrogen).

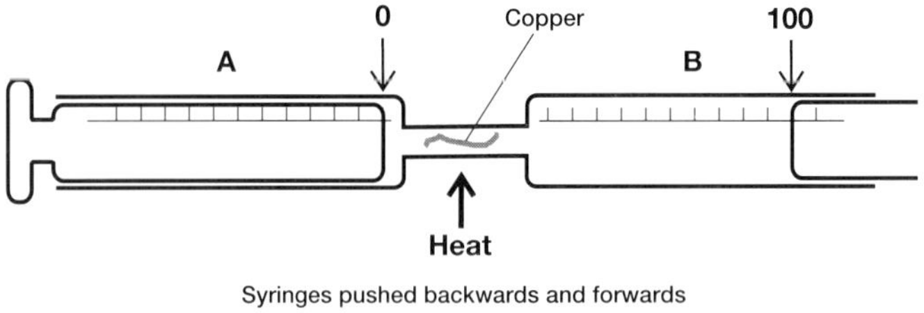


Fig. 5.5 *Experiment 3: forming an oxide*

**REMEMBER**

The whole apparatus **must** be allowed to cool down **before** volume readings are taken.

At the start: vol of air in syringe B = 100 $cm^3$
At the end: vol of air in syringe B = 80 $cm^3$
So, volume of air used up changing copper to copper oxide is 20 $cm^3$.

If the word equation for the reaction is:

COPPER (s) + PART OF AIR (g) ⟶ COPPER OXIDE (s)

The 'part of air' **must** be oxygen (remember our rules about chemical reactions).

Hence,

COPPER (s) + OXYGEN (g) ⟶ COPPER OXIDE (s)

**Presence of the noble gases**

| | | |
|---|---|---|
| Helium<br>Neon<br>Argon<br>Krypton<br>Xenon<br>Radon | All colourless gases which have no smell or colour | All present in air in small amounts (argon being the most abundant) |

The total amount of **all** the noble gases is about 1% of the air.

Chemical evidence of their presence is very difficult as these gases **do not react** to form new substances.

They used to be known as the **INERT** GASES (INERT – does not react) and in fact, no compounds of Helium, Neon and Argon have ever been found. Their name changed to NOBLE GASES after an English chemist was able to make Xenon and Krypton react for the first time in 1964.

## Noble gases – unreactive but useful

### ARGON

Used in light bulbs to provide an 'inert' gas which will not allow the **tungsten** filament to burn, even though it is white hot.

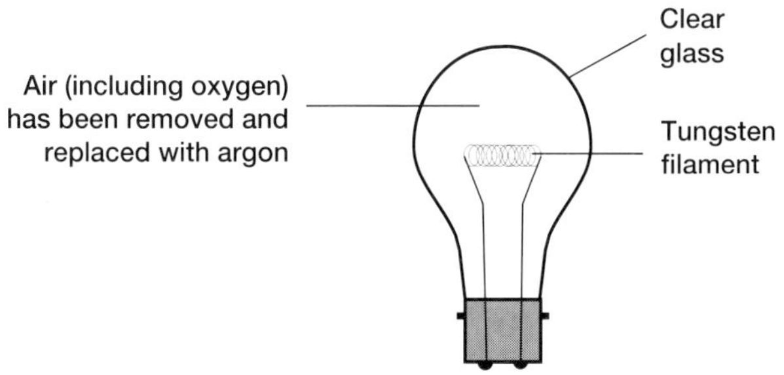


Fig. 5.6

### NEON

If a glass tube has the air removed and is filled with **NEON** instead, it will glow red when an electric current is put through the gas.

As glass tubes can be made into many shapes then **NEON SIGNS** of various designs can be made.

### HELIUM

This gas is much lighter than air and is a safer alternative to the explosive hydrogen as a gas to fill airships and balloons.

Because it boils at –269°C, it is used in liquid form as a coolant. At very low temperatures, the resistance to an electric current becomes so small, that materials become '**super conductors**'. Thus helium has definite uses in the electronics industry.

### KRYPTON

This has uses in **medicine** (used in lasers) and in **car maintenance** (engine tuning and wheel balancing machines).

Air is cooled to about –200°C. At that temperature, it will be a liquid. Water and carbon dioxide will have solidified during the cooling process and have been removed.

Liquid air is gently 'boiled' at about –190°C in a fractionating column.

At this temperature:

(i) Nitrogen will be a **gas** so it **rises** up the column where it is collected at the top.
(ii) Oxygen and noble gases are **still liquids** and so drip back down.

As with all fractionating columns, the top of the column is the coolest part as it is the furthest away from the 'heat'.

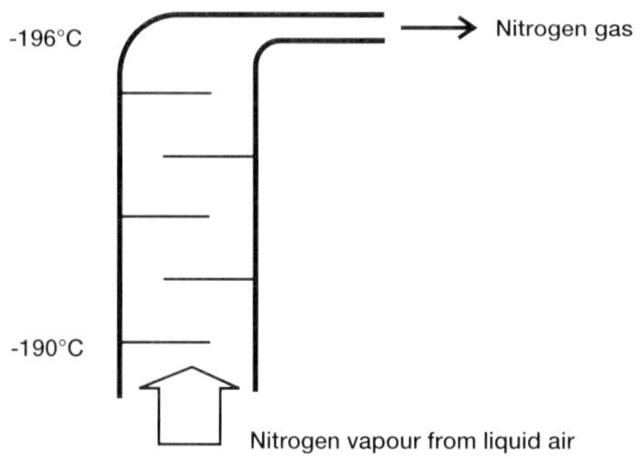


Fig. 5.8

## NITROGEN AS A 'CHEMICAL FEEDSTOCK'

'Chemical feedstock' is a term used by the chemical manufacturing industry to describe a substance which is a raw material from which other chemicals are made.

When we see the small bottles of chemicals in the laboratory, it is difficult to imagine that their presence is the tip of a very huge iceberg. In fact, around 3 and a half billion cubic metres of nitrogen are produced annually in the UK alone.

## PRODUCTION OF NITRIC ACID

When a large amount of energy (e.g. lightning in a thunderstorm), is applied to a mixture of oxygen and nitrogen, oxides of nitrogen can be formed. These oxides dissolve in water to form nitric acid.

We saw in Chapter 4 that reactions of bases with nitric acid produce salts which are called **nitrates.**

By far, the major use of nitric acid is in the manufacture of nitrates.

**FOR EXAMPLE**

Ammonium nitrate – fertiliser
Potassium nitrate – gunpowder
Silver nitrate – photography
Barium nitrate – explosives and fireworks ] These are made in much smaller quantities

**Nitro-compounds**
Nitric acid is used in the production of these compounds.

**FOR EXAMPLE**

Nitroglycerine – for explosives
Trinitrotoluene (TNT) – for explosives
Nitrocellulose – for explosives and paints

### PRODUCTION OF AMMONIA

Ammonia is a compound of nitrogen and hydrogen. It is a gas which has a very strong smell and is **very** soluble in water. A solution of ammonia in water is called **ammonium hydroxide** and this is alkaline.

Ammonia is made in vast quantities (over 80 million tonnes per year) and is used:

to make nitric acid
to make fertilisers (ammonium nitrate, ammonium sulphate, ammonium phosphate)
to make the liquid used in refrigerators.

As with most compounds of nitrogen, the chemistry of ammonia (gas) and its ammonious compounds, is quite complex and you will certainly study these in more detail later in your chemistry studies.

## Oxygen

- ☐ This is an **element**.
- ☐ Melting point –218°C; Boiling Point –183°C.
- ☐ Colourless gas at room temperature which has no smell.
- ☐ Reasonably soluble in water (think of fish!), to form a neutral solution.

- Reactive gas – responsible for burning, rusting, maintenance of life (respiration)
- Occurrence: approx 20% of the air (as oxygen gas)
  approx 45% of the earth's crust (as compounds)

## Chemical test to identify oxygen

Oxygen is the **only colourless gas which re-lights a glowing spill.**

### PREPARATION OF OXYGEN

**1. Thermal decomposition of potassium manganate (VII)***

This decomposes according to word equation:

| POTASSIUM MANGANATE (VII) (s) $\xrightarrow{\text{HEATING}}$ OXYGEN (g) + RESIDUE (s) |
|---|

You are not expected to name the residue at this stage!

**2. Decomposition of hydrogen peroxide**

Hydrogen peroxide is a liquid which can be used as bleach (i.e. to remove the colour from textiles or change dark hair into light-coloured hair). It is normally kept in dark bottles because in the light, hydrogen peroxide decomposes into oxygen and water. This is shown by the word equation:

| HYDROGEN PEROXIDE (l) $\longrightarrow$ OXYGEN (g) + WATER (l) |
|---|

This reaction is **very slow** – too slow for any practical purposes.

To speed up the reaction, we use a **CATALYST**, in this case, the catalyst used is **MANGANESE (IV) OXIDE**. This is a black powder which is insoluble in water.

When the reaction has finished, we can filter out the insoluble manganese (IV) oxide at the end. The black powder has speeded up the reaction, but has not **been changed** itself.

There are many examples of catalysts being used – some of them **slow reactions down.**

**THUS**

| A CATALYST IS A SUBSTANCE WHICH ALTERS THE RATE OF A CHEMICAL REACTION, BUT REMAINS CHEMICALLY UNCHANGED AT THE END OF THE REACTION |
|---|

* Potassium Permanganate

When writing word equations, it is usual to put the catalyst above the arrow

**THAT IS**

$$\text{HYDROGEN PEROXIDE (l)} \xrightarrow{\text{MANGANESE (IV) OXIDE}} \text{OXYGEN (g)} + \text{WATER (l)}$$

**OXYGEN AND BURNING**

We saw in chapter 3 that when magnesium burned in air, it **combined with oxygen** from the air to form magnesium oxide. As large quantities of energy were given out by this reaction, in the form of heat and light, then we say that the reaction is **exothermic**.

Also, the solid **gained mass** as it changed to magnesium oxide.

Contrast the reaction of magnesium and oxygen to the reaction of copper and oxygen – see the experiment in Fig. 5.5 (page 92).

Here, the copper **gains mass** as it combines with the oxygen of the air (just like magnesium). **But** there **is no burning.**

In fact, the reaction would seem to be endothermic, as heat energy is supplied to make the reaction happen.

In **both** cases, the combining reaction is called **OXIDATION**.

**THUS**

OXIDATION IS A CHEMICAL REACTION WHICH INVOLVES COMBINATION WITH OXYGEN

WHEN AN ELEMENT IS OXIDISED, ONLY ONE PRODUCT IS FORMED

BURNING (ALSO KNOWN AS COMBUSTION) IS AN EXOTHERMIC CHEMICAL REACTION WHICH INVOLVES COMBINATION WITH OXYGEN

## Burning compounds

We know that the burning of the candle in Fig. 5.3 (page 91) is an exothermic oxidation, but how do we know how many products are formed? Let us look at a modification of Fig. 5.3.

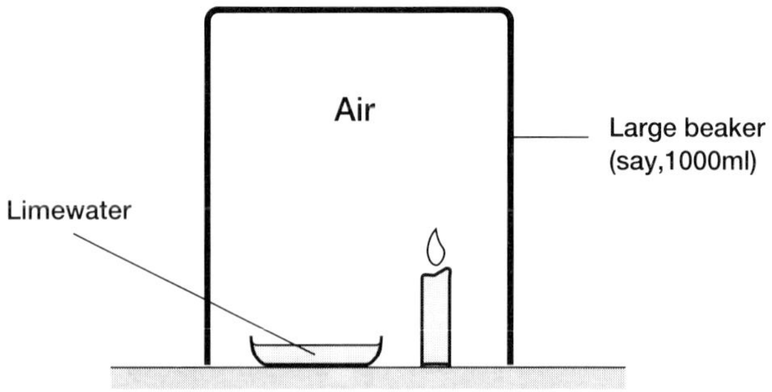


Fig. 5.9

1) We know that after a while, the candle goes out as the oxygen has been used up during the burning.
2) The beaker will mist up during burning. Sprinkling these droplets with anhydrous copper sulphate will, as they turn blue, tell you that **WATER** has been produced.
3) The limewater will become chalky, showing that **CARBON DIOXIDE** has been produced.
4) Black soot may well appear on the 'top' of the beaker – this is a sure sign that the burning – because of the limited oxygen supply – has been incomplete.

This experiment tells us that burning a candle produces at least two products (**CARBON DIOXIDE** and **WATER**).

So we **know** that a candle is **not** an element, but contains at least two substances.

| CANDLE (s) + OXYGEN (g) ⟶ WATER (g) (Hydrogen oxide) + CARBON DIOXIDE (g) |
|---|

The **hydrogen** in the water **MUST** have come from the candle. The **carbon** in the carbon dioxide **MUST** have come from the candle.

Looking at the word equation tells us that apart from the oxygen, **no other substances are involved,** so we say that:

| CANDLE WAX IS A COMPOUND WHICH CONTAINS HYDROGEN AND CARBON |
|---|

Compounds which are made up from **ONLY** the two elements **CARBON** and **HYDROGEN**, are called **HYDROCARBONS**.

**BEWARE!**

This experiment does not confirm the **exact** composition of candle wax, it merely tells you that it *contains* carbon and hydrogen.

## Burning glucose

If you lower some burning glucose into a large gas jar which has a little limewater on the bottom, you will also find that water and carbon dioxide are produced.

GLUCOSE (s) + OXYGEN (g) ⟶ WATER (g) + CARBON DIOXIDE (g)

You would be correct in assuming that glucose contains carbon and hydrogen. However as it is a **CARBOHYDRATE,** it will also contain **OXYGEN**.

It is not possible with these two experiments to say whether the oxygen in the products (carbon dioxide and water) came from the oxygen used in burning, or was present in the compound being burnt.

However, it **is** possible to say that:

**BURNING A COMPOUND PRODUCES MORE THAN ONE PRODUCT**

## Burning increases mass

As burning involves combination with oxygen,

**THE PRODUCT/S PRODUCED AS A RESULT OF BURNING ALWAYS HAS/HAVE MORE MASS THAN THE SUBSTANCE BEING BURNT**

Let us look at a few examples.

## Burning magnesium

MAGNESIUM (s) + OXYGEN (g) —BURNING⟶ MAGNESIUM OXIDE (g)

↑ Extra mass because of combined oxygen

This is a straightforward example as both the solid reactant and product can be weighed before and after the reaction.

### Lavoisier's experiments

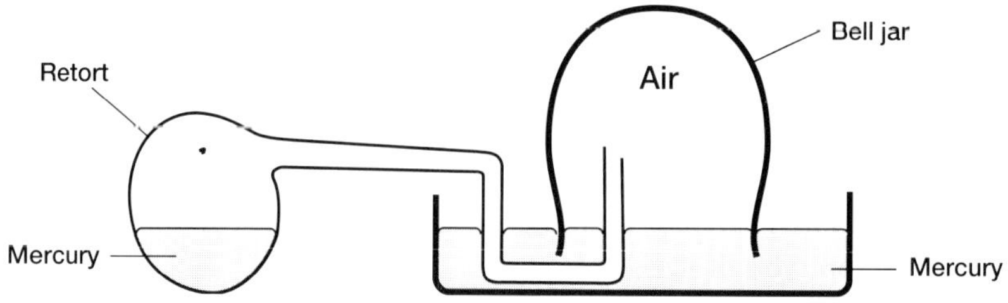


Fig. 5.10

1) He heated mercury in air and found that:
   (a) Red spots of mercury calx appeared on the surface of the mercury in the retort being heated.
   (b) The level of the mercury rose in the bell jar, showing that part of the air was 'used up'.
   (c) After about two weeks, there was no further rise of mercury in the bell jar. It was found that $1/5$ of the air had been used up in changing the mercury in the retort to mercury calx.
2) He scraped the red calx from the surface of the mercury and heated this in a separate retort. He found that:
   (a) The volume of gas produced by heating the red calx was the **same volume** of gas that had been 'used up' in the first part of the experiment when the calx had been formed.
   (b) The gas produced by heating the calx was the **same gas** which Priestley had discovered and called dephlogistigated air.
3) Lavoisier burned other substances in this gas, e.g. carbon and sulphur and found that the oxides formed, dissolved in water to form **acids**. He gave this gas the name **OXYGEN** (acid producer).

### DEATH OF THE PHLOGISTON THEORY

1) Lavoisier concluded from his experiments that when things burn, they combine with oxygen from the air and so **gain** mass rather than lose it as phlogiston.
2) The introduction of the **chemical balance** and accurate weighing of reactants, before burning and products, after burning (as in the formation of magnesium oxide), proved beyond all doubt that burning is a **combining** reaction – rather than a thermal decomposition, with the loss of the nebulous phlogiston.

### DEATH OF A GREAT CHEMIST

Lavoisier was one of the world's great chemists and his work contributed much to a better understanding of the subject. Sadly, he lived at the time of the French Revolution and he had two serious disadvantages working against him.

Firstly, he was a member of a wealthy family and secondly, he was a tax collector. One of these would have been bad enough in the disturbed climate of those times. The two together were too much for those who ruled in those days and so his life was brought to an end by the guillotine in May 1794.

## Oxygen and fuels

Substances which we burn with the express purpose of using the energy (heat, light) which is released by the exothermic reaction with oxygen, are called **FUELS.**

Fuels come in **three** main categories:

(1) **LIVING-FUELS: wood** and materials from plants.

(2) **SEMI-FOSSIL FUELS:** e.g. **peat,** which is compressed plant material that is found in peat bogs. The material is cut out, dried and burned.

(3) **FOSSIL FUELS: coal, oil, gas.**

**All fuels** have the following in common:

1) **Fuels by themselves do not release energy. It is the FUEL/OXYGEN REACTION which releases energy.**

2) Fuels **contain carbon compounds.** This means that when they burn, **carbon dioxide** is released into the atmosphere.

### FOSSIL FUELS

It must be realised that it took over 100 million years to **produce** a lump of coal which can be burned on a fire in your home in about **half an hour.**

The fact of the matter is, that we are **using fossil fuels faster than they can be replaced.** At some stage they will run out.

Various people and organisations have different predictions as to when this will be. We should not be gloomy about the situation, but we **should be aware** of the possibilities.

At present, a great deal of research is being done, all over the world, to:

1) find alternative ways of generating electricity, so as to reduce the demand for coal which is being burnt in power stations in vast quantities
2) find ways of using our existing oil supplies more economically by developing alternative materials to those that at present come from oil e.g. plastics, petrol for cars.

**COAL**

This is the fossilised remains of plant material which grew on the earth over 100 million years ago. As the trees died and fell into the swamps, they were covered by layers of mud which kept the oxygen away and so prevented the organisms which cause decay from respiring. Over the years, as further layers of rock were laid on top, heat and pressure changed the plant material into the hard rock, coal, which we know today.

Coal often contains **sulphur compounds** as well as carbon compounds. This means that when coal is burnt, both gases, **carbon dioxide** and **sulphur dioxide** will be given off.

Coal also contains a great variety of chemicals (some being hydrocarbon) which can be obtained from **coal tar** and there are many examples of plants being set up to extract these chemicals from coal.

When coal is heated in a retort, then inflammable gas (town gas) and coal tar are given off. What remains is almost pure carbon (coke). Nothing is wasted!

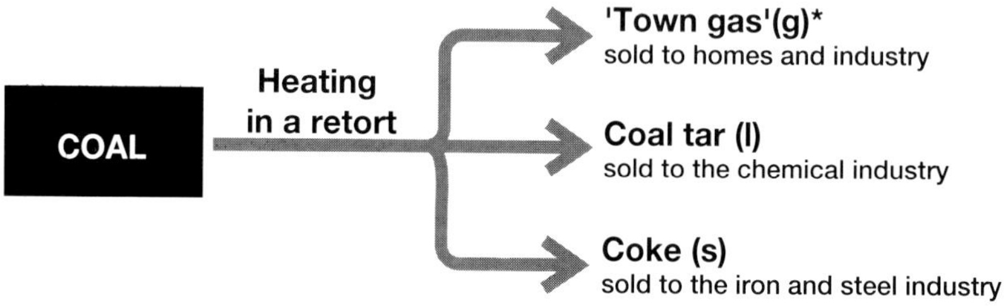


*This used to be the gas supplied to our homes and bunsen burners before Natural gas (methane) was piped from the North sea

Fig. 5.11

# Air pollution caused by burning coal

## 1. CARBON DIOXIDE

Whilst carbon dioxide is a harmless gas in itself, its increase in the atmosphere has been partly responsible for the concern which we share about **global warming** which is caused by the **greenhouse effect.**

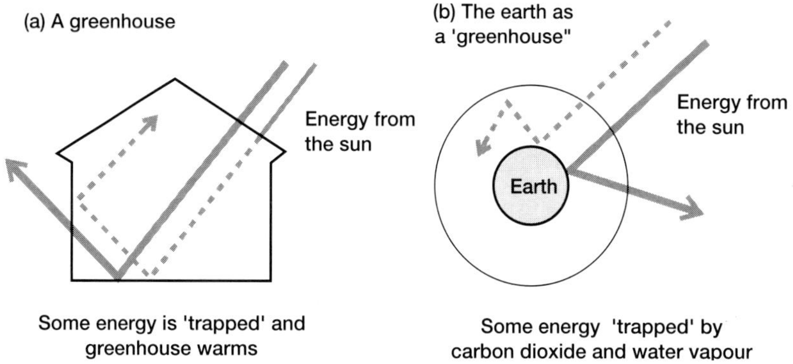


Fig. 5.12 *The greenhouse effect*

Energy from the sun enters a greenhouse (see (a)). As it passes through the glass, some of it **changes wavelength**, which means that it is no longer able to escape out of the glass. So the greenhouse becomes warmer. The atmosphere around the Earth acts like the glass of a greenhouse. Energy reaches the earth and most of it is reflected back into space. However, the atmosphere contains certain gases (called the **greenhouse gases)** which absorb some of the energy and so the atmosphere warms. As the atmosphere warms, so does the Earth, as there is a constant stream of energy reaching it from the sun and only **part** of this is radiated away from the atmosphere to space.

**Greenhouse gases** which are responsible for global warming include:

**carbon dioxide, water vapour, methane, nitrogen oxides.**

## 2. SULPHUR DIOXIDE

Many substances on the earth are sulphur compounds. When these are heated, they decompose to release sulphur dioxide. Burning coal also releases sulphur dioxide into the air.

| |
|---|
| SULPHUR DIOXIDE DISSOLVES IN THE WATER OF THE ATMOSPHERE TO FORM **ACID RAIN** |

**REMEMBER:**

'Normal rain' is always *slightly* acidic, because of the carbon dioxide which dissolves in the water of the atmosphere. It is, in fact, a solution of **carbonic acid** which is a weak acid. The **acid rain** produced because of the dissolved sulphur dioxide has a **much lower pH than normal rain** and it causes serious problems for wildlife and the environment such as:

1) direct damage to trees and plants;
2) releasing aluminium which is normally 'held' in neutral soils (this will find its way into streams and rivers causing serious problems for the fish);
3) making the soil **too** acid (certain fertilisers do not release their 'goodies' if the soil is too acid, resulting in poor crops);
4) damage to buildings and man-made construction (we have already seen about the action of an acid on limestone – calcium carbonate).

## Reducing sulphur emissions from chimneys

Much time and money is currently being spent trying to solve this problem.

One solution is to pass the smoke containing the sulphur dioxide through a mist of water which then trickles over lime. The sulphur dioxide dissolves in the water and the resulting acid reacts with the base (calcium oxide) to produce the neutral salt, calcium sulphate and water.

i) SULPHUR DIOXIDE (g) + WATER (l) ⟶ SULPHURIC ACID (l)

ii) SULPHURIC ACID + CALCIUM HYDROXIDE ⟶ CALCIUM SULPHATE + WATER

The calcium sulphate is not just thrown away – but is sold to the building industry which uses it as one of the ingredients of plaster.

### SOOT

Soot is made up of black specks of unburnt carbon. They occur when burning is not complete – maybe because of a restricted supply of oxygen.

In the 1950s, soot from coal fires mixed with the London winter fogs to cause **smog** (a smoke/fog mixture). This caused enormous problems – not only for traffic, as visibility was severely reduced, but also for those who suffered with chest infections.

A **clean air** programme was set up and certain areas were known as **smokeless zones**. If you lived – or live, in a smokeless zone, then you would not be allowed to burn 'household' coal on your fire. You would have to use **smokeless coal** which is mainly coke, instead.

## Oil and gas

These fuels consist **only** of hydrocarbons and, as they were formed in the same way and are generally found close to each other, we shall deal with them together.

Oil was formed in the ancient seas over 100 million years ago from the little animals which lived there. As they died, so they sunk to the bottom and mixed with the mud to form a 'zoological ooze'. As with coal, oxygen was excluded from this ooze and so the decaying organisms were not able to do their stuff. As the waters receded and more and more layers of rock were laid down on top of the ooze, temperature, pressure and time changed the ooze into the oil and gas which we pump out and use today.

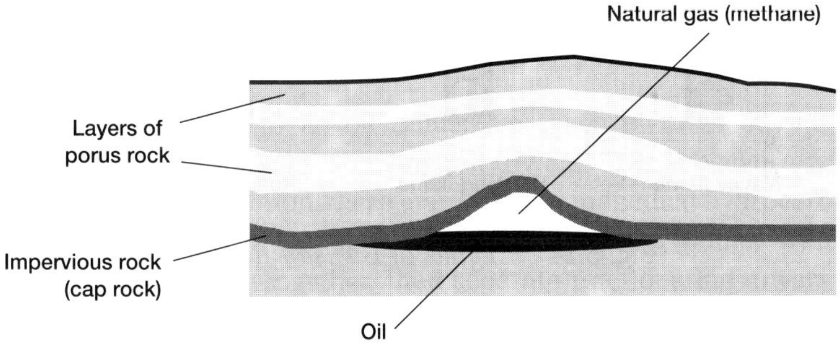


Fig. 5.13 *Location of oil and gas*

**BURNING GAS**

Natural gas is the hydrocarbon **METHANE**, which burns to produce carbon dioxide and water.

| |
|---|
| NATURAL GAS (g) + OXYGEN (g) ⟶ CARBON DIOXIDE (g) + WATER (l)<br>(Methane) |

If your laboratory is connected to the gas mains supply, then this is the gas that you will use as you use your Bunsen burners.

## AIR POLLUTION CAUSED BY BURNING OF OIL

Nearly all forms of transport burn some form of hydrocarbon as a fuel. This means that enormous quantities of carbon dioxide and water vapour are released into the air – thus adding to the amount of greenhouse gas in the atmosphere.

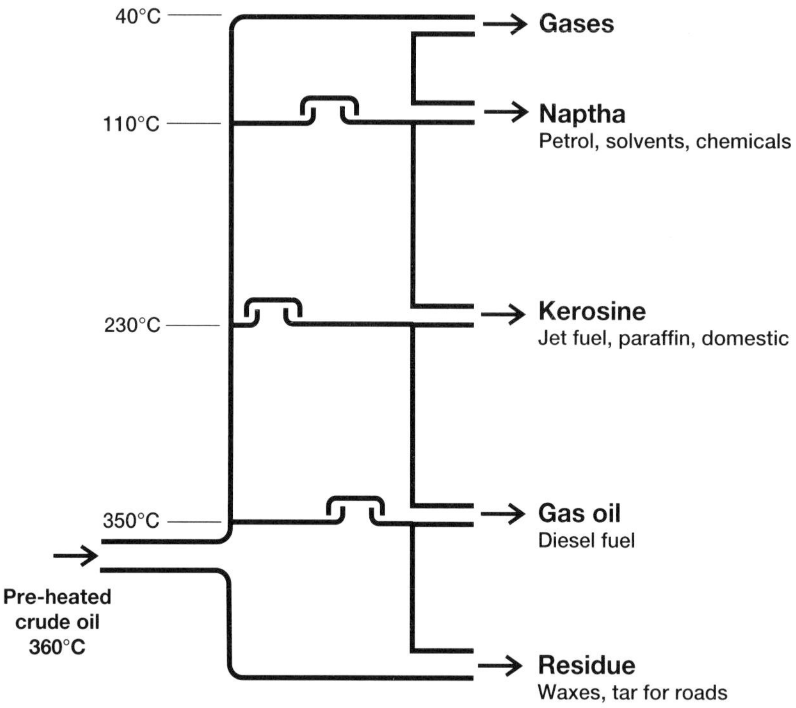


Fig. 5.15

### Pollution from motor cars

Apart from carbon dioxide and water, a motor car will also emit the following gases from its exhaust pipe:

1) nitrogen oxide
2) carbon monoxide

Both gases are poisonous.

**Nitrogen oxide** is formed when the spark and subsequent high temperatures (caused by the petrol-air mixtures exploding) cause the nitrogen and oxygen, which is present in the air in the cylinders, to combine.

**Carbon monoxide** is formed by the incomplete burning of the hydrocarbon (petrol) because of the limited amount of oxygen available. As carbon monoxide leaves the exhaust pipe, it combines with the oxygen of the air to form carbon dioxide.

| CARBON MONOXIDE (g) | + | OXYGEN (g) | ⟶ | CARBON DIOXIDE (g) |
|---|---|---|---|---|
| ↑ | | ↑ | | ↑ |
| Poisonous gas from exhaust | | From air | | Non-poisonous gas |

As carbon monoxide is always part of a car's exhaust emission, it is only safe to have the engine running where there is a good supply of oxygen. This is why engines are **NEVER** left running in garages with the garage door closed. Otherwise the available oxygen is soon used up and the amount of carbon monoxide in the air rises to dangerous levels. (See BIOLOGY for details about carbon monoxide poisoning).

Emissions of nitrogen oxide and carbon monoxide in car exhausts have been reduced by passing the gases through a **CATALYTIC CONVERTER** which changes the poisonous gases into safer ones.

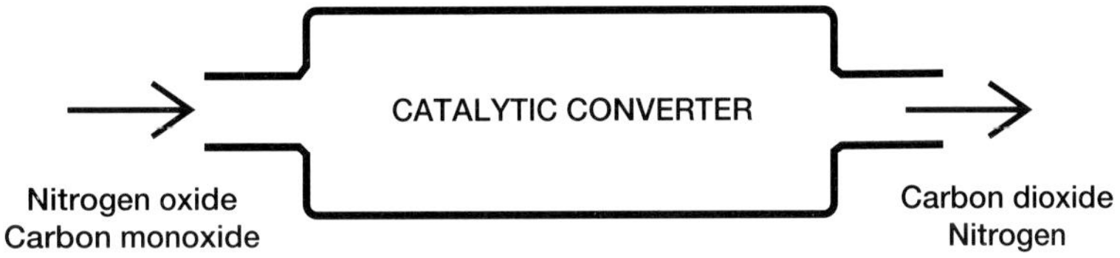


Fig. 5.16

## Oxygen and rusting

The term rusting refers **only** to the **corrosion** of **iron.**

**Corrosion** is a process which chemically alters metals – usually because of the action of substances present in the air.

These substances react with the metals to form powdered compounds on the surface. These powders flake away exposing more metal to the corrosive substances. This newly-exposed surface forms more powdered compounds and so on, until the original metal is 'eaten away'.

Rusting is one of the most expensive chemical reactions in the world.

Iron is used in so many ways: e.g. transport, buildings, civil engineering, farm and garden tools, fence posts and fencing wire … the list is endless!

(c) What does the action of dilute acid on the mineral A tell you about the chemical nature of the mineral?

(d) What type of compound was:

(i) the powder

(ii) the filtrate?

(e) What was the gas produced when acid was added to mineral A? Give a reason.

14. The apparatus below was used to investigate the products formed when various fuels are burned.

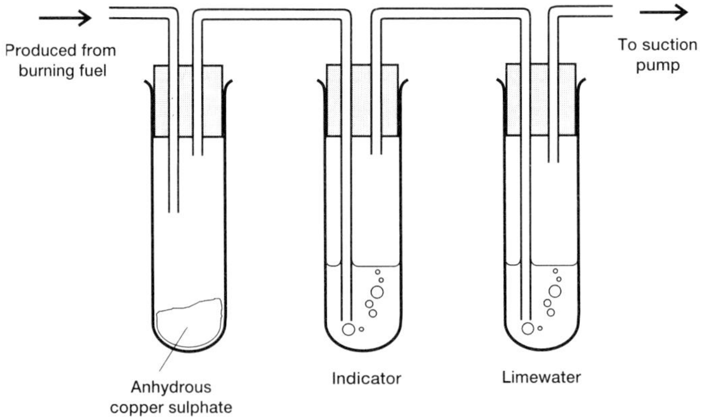


(a) What substance would be detected by:

i) the anhydrous copper sulphate

ii) the limewater

In each case, say what you would see happening.

(b) The table shows the results obtained from three fuels, A, B and C.

| Fuel | Anhydrous copper sulphate | Indicator | Limewater |
|---|---|---|---|
| A (solid) | no reaction | 6 | turns milky |
| B (liquid) | turns blue | 3 | turns milky |
| C (gas) | turns blue | 7 | no reaction |

i) Which fuel contained no carbon? Give a reason.

ii) What substance might the fuel you have named in i) be? Explain your choice.

# Further questions on ...

15. Crude oil is practically distilled to produce a range of fractions which differ from each other in important ways.

    Explain, giving your reasons:
    (a) whether crude oil is a mixture or a compound;
    (b) why fractional, rather than simple, distillation is used;
    (c) the everyday use of at least three of the fractions and why they are used for that purpose.

16. (a) When petrol burns in a car engine two products are formed in large amounts. What are they? How would you test for these substances in the laboratory?
    (b) It is also possible for three other substances to be formed in car engines: carbon monoxide, sulphur dioxide and nitrogen dioxide.

    How are these substances produced and how do they give rise to environmental problems?

17. Why is it possible to suspend a match head in a Bunsen flame without the match igniting?

18. Below is a diagram of the 'roaring' Bunsen flame.

(a) Explain why the 'roaring' noise is produced.

(b) Why is the inner part of the flame cone-shaped?

(c) Why is the inner cone a lighter blue colour than the rest of the flame?

## ... gases

**TABLE 6.1**

| Element | Is it shiny? | Is it bendy? | Does it conduct electricity? | Density g/cm$^3$ |
|---|---|---|---|---|
| Aluminium | ✓ | ✓ | ✓ | 2.7 |
| Calcium | ✓ | ✓ | ✓ | 1.6 |
| Carbon | × | × | ✓ | 2.2 |
| Copper | ✓ | ✓ | ✓ | 8.9 |
| Iron | ✓ | ✓ | ✓ | 7.9 |
| Lead | ✓ | ✓ | ✓ | 11.3 |
| Magnesium | ✓ | ✓ | ✓ | 1.7 |
| Mercury | ✓ | (liquid) | ✓ | 13.6 |
| Potassium | ✓ | ✓ | ✓ | 0.9 |
| Sodium | ✓ | ✓ | ✓ | 1.0 |
| Sulphur | × | × | × | 2.1 |
| Zinc | ✓ | ✓ | ✓ | 7.1 |

Now we use 'Best of three ticks' (i.e. 2 or more) to sort them into two groups. Ticks on the left, crosses on the right.

**TABLE 6.2**

| ✓✓✓ (or ✓✓×) | ××× or (××✓) |
|---|---|
| Aluminium, calcium, copper, iron, lead, magnesium, mercury, potassium, sodium, zinc. | Carbon, sulphur |

↑ THIS GROUP ARE **METALS**
1) all are shiny
2) all are bendy
3) all conduct electricity

↑ THIS GROUP ARE **NON-METALS**
1) all are dull
2) all are brittle
3) nearly all do not conduct electricity (carbon is the exception)

Having sorted them into two groups, metals and non-metals, we notice a great clump of elements in the metals and only two in the non-metals.

This is very much in line with the periodic table where there are about 80 metals and about 20 non-metals.

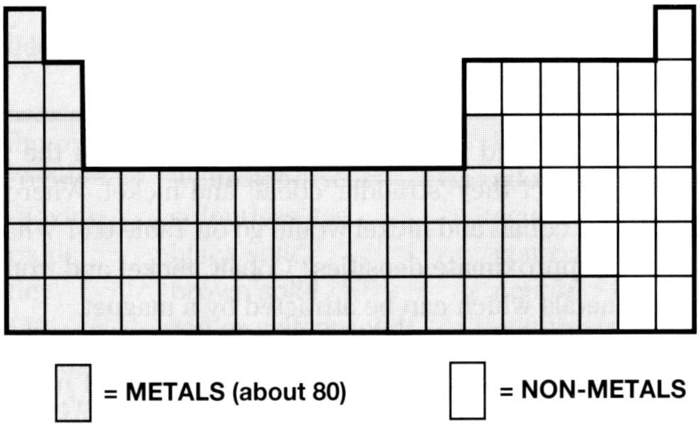


Fig. 6.1

Using **densities,** we can sort our elements into smaller groups.

Use two **'density lines'** – one for 10 g/cm$^3$, one for 2.5 g/cm$^3$. These lines run horizontal across the page.

## Placing the element

1) Is it a metal or non-metal (see previous table)
2) What is its density? Place it above/below the appropriate density line.

**TABLE 6.3**

| | Metal | Non-metal | |
|---|---|---|---|
| | Lead (11.3) | | |
| | Mercury (13.6) | | |
| 10 g/cm3 | - - - - - - - - - - | - - - - - - - - - - | 10 g/cm$^3$ |
| | Copper (8.9) | | |
| | Iron (7.9) | | |
| | Zinc (7.1) | | |
| | Aluminium (2.7) | | |
| 2.5 g/cm$^3$ | ══════════ | ══════════ | 2.5 g/cm$^3$ |
| | Calcium (1.6) | Carbon (2.2) | |
| | Magnesium (1.7) | Sulphur (2.1) | |
| | Sodium (1.0) | | |
| | Potassium (0.9) | | |

## Looking at the solubility of oxides

| | Metal oxides | Non-metal oxides |
|---|---|---|
| OXIDES **INSOLUBLE** IN WATER | These bases will neutralise acids to form **salts** | |
| OXIDES **SOLUBLE** IN WATER | These dissolve in water to form **ALKALIS** | These dissolve in water to form **ACIDS** |

## Bases and Alkalis

**ALL** metal oxides are solids called **BASES**

**SOME** bases dissolve in water to form **ALKALIS**

**FOR EXAMPLE**

Calcium / Magnesium / Sodium } Oxide + Water ⟶ Calcium / Magnesium / Sodium } Hydroxide

Alkalis and bases will neutralise acids to form salts

**FOR EXAMPLE**

1)

| | | | | | | |
|---|---|---|---|---|---|---|
| SODIUM HYDROXIDE (Alkali) | + | SULPHURIC ACID (Acid) | ⟶ | SODIUM SULPHATE (Salt) | + | WATER (Water) |

2)

| | | | | | | |
|---|---|---|---|---|---|---|
| COPPER OXIDE (Alkali) | + | SULPHURIC ACID (Acid) | ⟶ | COPPER SULPHATE (Salt) | + | WATER (Water) |

## Acidic non-metal oxides

Let us look again at the oxides of carbon and sulphur.

Both oxides (in the case dioxides) are colourless gases which dissolve in water to form acids.

CARBON DIOXIDE (g) + WATER (l) ⟶ CARBONIC ACID (l)

SULPHUR DIOXIDE (g) + WATER (l) ⟶ SULPHURIC ACID (l)

Both are acids, but testing with a universal indicator shows that they are of different strengths.

| Acid | Colour with indicator | pH | Strength of acid |
|---|---|---|---|
| Carbonic acid | Yellow/Orange | 5 | Weak |
| Sulphuric acid | Red | Less than 4 | Strong |

The small amount of carbon dioxide in the air will dissolve in atmospheric water and this is why rainwater is *always slightly acid.*

However, in certain industrial locations, sulphur dioxide is emitted in reasonably large quantities. In these locations the sulphur dioxide dissolves in the atmospheric water to produce **acid rain**.

It is this particular form of rain which, as we have seen, is harmful to the environment.

## Another look at the limewater test

**FACTS**

1) Limewater is a solution of calcium hydroxide in water
2) Carbon dioxide dissolves in water to form carbonic acid.

When you bubble carbon dioxide into limewater, it dissolves in the water to form carbonic acid – see **A**.

The carbonic acid reacts with the calcium hydroxide to form the insoluble salt, calcium carbonate **B**.

**A** CARBON DIOXIDE (g) + WATER (l) ⟶ CARBONIC ACID (l)

**B** CARBONIC ACID (l) (Acid) + CALCIUM HYDROXIDE (l) (Alkali) ⟶ CALCIUM CARBONATE (s) (Salt) + WATER (l) (Water)

**If carbon dioxide continues to bubble through, the solution clears.**

When all the calcium hydroxide has been changed into calcium carbonate (the chalky precipitate), then the carbon dioxide will dissolve in the water to form more carbonic acid – see **C**.

This carbonic acid reacts with the calcium carbonate to produce a soluble salt called calcium hydrogen carbonate – see **D**.

As calcium hydrogen carbonate is soluble – the mixture clears.

| | | | | | | | | | |
|---|---|---|---|---|---|---|---|---|---|
| **C** | CALCIUM CARBONATE | + | WATER | + | CARBON DIOXIDE | ⟶ | CALCIUM CARBONATE | + | CARBONIC ACID |

| | | | | | | | | |
|---|---|---|---|---|---|---|---|---|
| **D** | CALCIUM CARBONATE | + | CARBONIC ACID | ⟶ | CALCIUM HYDROGEN CARBONATE | + | WATER |
| | (Base) | | (Acid) | | (Soluble salt) | | (Water) |

# What do you know about ...

1. Make a table with three headings;

   Metallic element
   Non-metallic element
   Compound

   Put the following substances into the correct parts of your table:

   | | |
   |---|---|
   | carbon | copper sulphate |
   | iron filings | magnesium |
   | mercury | oxygen |
   | sodium | sulphur |
   | water | zinc |

2. Select the 'odd one out' in each of the following groups, giving a reason for your choice.

   (a) copper, iron, mercury, zinc
   (b) copper, iron, lead, sodium
   (c) carbon dioxide, copper oxide, nitrogen oxide, sulphur dioxide
   (d) carbon, iron, sulphur, zinc

3. Using the information given in the table below, say whether the elements A, B, C, D, E and F are metals or non-metals. You do not have to identify the elements.

   | Element | Physical state | Oxide formed when burned in air | Solubility |
   |---|---|---|---|
   | A | Solid | White oxide | soluble (pH 12) |
   | B | Solid | Black oxide | insoluble |
   | C | Solid | White oxide | soluble (pH 4) |
   | D | Solid | Gaseous oxide | soluble (pH 5) |
   | E | Liquid | Lead oxide | insoluble |
   | F | Gas | Liquid oxide | (pH 7) |

4. Look at the oxides in Question 3 and say which of the elements form oxides which are bases, and which of the elements form oxides which are alkalis. Give your reasons for your reactions.

## ... metals and non-metals?

# What do you know about ...

5. Magnesium, potassium manganate (VII), copper and sulphur were all heated in air. Sadly, when the results were written down, they were identified as A, B, C, D. Look at the results below and then answer the questions which follow.

| Substance | Result of heating |
|---|---|
| A | The yellow solid melted to give a brown liquid. This liquid burnt with a blue flame to form a gas which turned indicator red. |
| B | The purple crystals changed to a black powder. A gas was given off which relit a glowing spill. |
| C | The shiny ribbon burned with a bright white light. A white solid forms. |
| D | The pink shiny form glows red hot in the Bunsen flame. On cooling, a black coating can be seen on the foil. |

(a) Apart from the list of names in the original list, what evidence is provided here to show that one of the substances heated is a compound?
(b) Give the names of A, B, C and D. Give reasons for your selection.
(c) The product formed when magnesium was burnt in air was dissolved in water and some universal indicator was added. What would you expect to happen and why?
(d) (i) What two pieces of information in the table tell you that sulphur is a non-metal?
(ii) Describe other experiments you could do to prove that sulphur is a non-metal.
(e) What does this table tell you about the comparative reactivities of copper and magnesium?

6. (a) Discuss the physical differences between metals and non-metals.
(b) Discuss the chemical differences between metals and non-metals.
(c) Mention any properties of particular metals or non-metals that do not fit the usual pattern you have discussed in (a) and (b).
(d) Described how you would make some magnesium oxide and some carbon dioxide in the laboratory. What physical and chemical differences do these two oxides have?

## ... metals and non-metals?

# What do you know about ...

7. (a) A substance (A) burns in air with a blue flame to form an acidic oxide. A can be changed into a black solid which conducts electricity. What, giving reasons, can you say about A?
   (b) B conducts electricity and combines with oxygen to form a solid which is insoluble in water. What, giving reasons, can you say about B?

8. Carbon, hydrogen, mercury, oxygen, sodium

   (a) Which of the elements listed would conduct electricity?
   (b) Which two of the elements listed would react together to form a compound which will turn a universal indicator blue? Give a reason.
   (c) Which two of the elements listed, would react together to form a compound which will be neutral to a universal indicator? Give a reason.
   (d) Which two of the elements listed above would react together to form a compound which will turn a universal indicator orange? Give a reason.

9. P, Q and R are three different elements. Use the information in the table below to answer questions about the elements and their oxides.

| Substance | M.P. °C | B.P. °C | Density $g/m^3$ | Effect on acid | Effect on alkali |
|---|---|---|---|---|---|
| P | –259 | –252 | 0.00009 | no effect | no effect |
| Oxide of P | 0 | 100 | 1.0 | no effect | no effect |
| Q | 650 | 1100 | 1.7 | reacts rapidly | no effect |
| Oxide of Q | 1780 | 2540 | 1.5 | neutralises acid | no effect |
| R | sublimes | 4200 | 2.3 | no effect | no effect |
| Oxide of R | sublimes | –78 | 0.002 | no effect | neutralises alkali |

   (a) At room temperature, which of the above substances is/are:
      (i) solid/s
      (ii) liquid/s
      (iii) gas/es?

... metals and non-metals?

**BURNING METALS**

This test will only be a rough guide as we can only use arbitrary criteria such as:

1) how brightly does the metal burn?
2) how fiercely does the metal burn?

Even so, by using this test it is possible to put some of the metals into an order of reactivity.

**FOR EXAMPLE**

| | |
|---|---|
| Sodium/Potassium | React with oxygen without burning |
| Calcium | Brilliant fierce red flame |
| Magnesium | Brilliant white flame |
| Zinc | Bright, fierce, blue/green flame |
| Iron | Yellow sparks |
| Copper/Lead/Tin/Gold | Do not burn at all |

Aluminium can be shown to burn fiercely, but this is quite difficult to do in the laboratory as aluminium quickly becomes coated with a film of aluminium oxide which prevents any further reaction. Mercury is **NEVER** heated in a laboratory.

## Reacting metals with water

Metals that do so, react with water according to the word equation:

$$\text{METAL (s)} + \text{WATER (l)} \longrightarrow \text{METAL HYDROXIDE (aq)} + \text{HYDROGEN (g)}$$

This means that when metal/water reactions are being investigated, we should look for:

1) bubbles of gas (hydrogen) being given off;
2) formation of metal hydroxide – indicator shows alkali is present.

In each case below, a small sample of metal was put into a dish (in the case of potassium and sodium) or test tube which contained water that had some universal indicator added.

**TABLE 7.1**

| Metal | Action with cold water |
|---|---|
| Potassium | Whizzes around, bursts into purple flame, burning any hydrogen as it is produced. Colour of water changes to purple as **potassium hydroxide** is produced. |
| Sodium | Whizzes around, becomes hot enough to melt itself as it reacts. Hydrogen is given off. Colour of water changes to purple as **sodium hydroxide** is produced. |
| Calcium | Sinks, then rapid bubbles of hydrogen appear. Colour of water changes to purple as **calcium hydroxide** is produced. As calcium hydroxide is not very soluble it appears as a solid precipitate in the water. |
| Magnesium, Zinc, Iron, Copper | These reactions occur very slowly in cold water – results could take days! The reactions are speeded up by using steam instead of cold water. |

## Reacting metals with steam

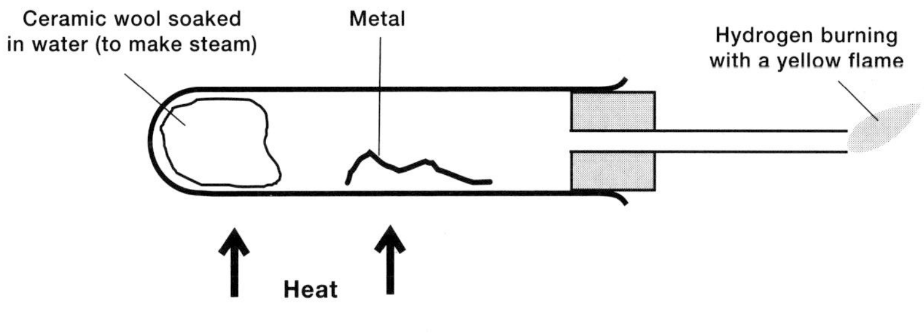


Fig. 7.1

Metals which react with steam do so according to the word equation:

METAL (s) + STEAM (g) ⟶ METAL OXIDE (s) + HYDROGEN (g)

This time we look for:

1) production of the metal oxide;
2) production of hydrogen – this will burn at the end of the tube.

**TABLE 7.2**

| Metal | Action with steam |
|---|---|
| Magnesium | A bright white light in the test tube (heat often enough to soften the glass of the test tube) and a strong, long hydrogen flame produced. |
| Zinc | A glow spreads throughout the zinc powder, which turns yellow (white when cool). A medium hydrogen flame is produced. |
| Iron | A dull red glow spreads throughout the iron and a tiny hydrogen flame is produced. |
| Copper | Nothing happens in the test tube – no hydrogen flame. **NO REACTION** |

## COMPOSITION OF WATER

Look again at Fig. 7.1. We use this apparatus to tell us what water is made of.

Consider the magnesium/steam reaction:

MAGNESIUM (s) + STEAM (g) ⟶ MAGNESIUM OXIDE (s) + HYDROGEN (g)

Look at the products:

1) The **oxygen** in magnesium oxide can **only** have come from the water.
2) The **hydrogen** gas can **only** have come from the water.
3) The **magnesium** in magnesium oxide is one of the original reactants.

*NB: The amount of oxygen already present in the test tube would be so small as to have no significant effect on the results of this reaction.*

This means that water contains the elements hydrogen and oxygen and nothing else.

We sometimes call water **hydrogen oxide** ($H_2O$). [Later, you will find that water can also be referred to as hydrogen hydroxide (HOH).]

**TABLE 7.3**

| Metal | Action of Water | Action of Steam |
|---|---|---|
| Potassium, Sodium, Calcium | React to form **metal hydroxide** and **hydrogen** | Too violent and dangerous! |
| Magnesium, Zinc, Iron | Very slow reaction | React to form **metal oxide** and **hydrogen** |
| Copper | No reaction | No reaction |

## POSITION OF HYDROGEN IN THE REACTIVITY SERIES

Although hydrogen is not a metal (it does not conduct electricity, is not shiny or bendy), it often behaves in a **chemical way** like a metal.

This is why it is not put in the main body of the periodic table. However, it is the lightest element and it is put **near** the metals, just outside the main body at the top – sometimes at the top left – sometimes in the middle.

The reactions with steam and water will give us a clue where hydrogen goes.

Look again at the reaction between steam (hydrogen oxide) and magnesium:

**Magnesium + Hydrogen oxide ⟶ Magnesium oxide + Hydrogen**

Fig. 7.2

In this case, magnesium has **DISPLACED** hydrogen from water.

Another way of looking at the reaction is to say that magnesium has **taken the oxygen away** from hydrogen oxide and has combined with the oxygen (i.e. been **oxidised**) to form magnesium oxide.

**The process of taking oxygen away by another** substance is called **REDUCTION**.

The substance which **does the taking away**, is called a **REDUCING AGENT**.

**SO WE SAY:**

Magnesium is the **reducing agent** which **reduces water** (hydrogen oxide) to **hydrogen.**

Note that, in this reaction, magnesium has itself combined with oxygen – it has been oxidised.

**OXIDATION – THE ADDITION OF OXYGEN TO A SUBSTANCE**

**REDUCTION – THE REMOVAL OF OXYGEN BY ANOTHER SUBSTANCE**

If we look at Table 7.3, we can see that:

1) Iron is able to reduce water to hydrogen – so Iron is **higher** than hydrogen in the reactivity series.
2) Copper is not able to reduce water to hydrogen – so copper is **lower** than hydrogen in the reactivity series.

This means that we can place hydrogen in reactivity series below iron and above copper.

**THUS**

ZINC
IRON
HYDROGEN
COPPER

## Reacting metals with dilute acids

We have seen before that when metals react with acids, they produce a salt and hydrogen.

**FOR EXAMPLE**

(i)

METAL + HYDROCHLORIC ACID ⟶ METAL CHLORIDE + HYDROGEN

(ii)

METAL + SULPHURIC ACID ⟶ METAL SULPHATE + HYDROGEN

Notice that the reaction of a metal with a dilute acid is a **DISPLACEMENT REACTION**.

The metal **displaces hydrogen** from the acid to form a salt.

## AN EVEN FURTHER LOOK AT ACIDS

**TABLE 7.4**

| Common name of acid | Chemical name of acid | Salts formed |
|---|---|---|
| Hydrochloric acid | Hydrogen chloride | — Chlorides |
| Nitric acid | Hydrogen nitrate | — Nitrates |
| Sulphuric acid | Hydrogen sulphate | — Sulphates |
| Carbonic acid | Hydrogen carbonate | — Hydrogen carbonates |

**All** the acids are **hydrogen** compounds.

**SO**

**AN ACID IS A COMPOUND WHICH CONTAINS HYDROGEN THAT CAN BE REPLACED BY A METAL WHICH IS HIGHER THAN HYDROGEN, IN THE REACTIVITY SERIES**

H is the chemical symbol for hydrogen. Thus it is logical that the pH scale, which measures the strength of acids, relates to the amount of hydrogen present. This was discovered in 1909 by a French chemist called Sorenson who referred to *poissance d'Hydrogene* – or pH.

[The hydrogen present is not in thc form of a gas ($H_2$), but in the form of **hydrogen ions** ($H^+$).]

## SOME REACTIONS OF METAL WITH ACIDS

### 1. Sodium and potassium

These metals are **NEVER** reacted with acids. The reaction would be too violent and dangerous.

After considering reactions of metals with water, steam, acids and the salts of other metals, we have now been able to produce a reactivity series which includes hydrogen.

We saw that magnesium was able to reduce water (hydrogen oxide) to hydrogen **because magnesium was higher in the reactivity series than hydrogen.**

Thus we can use the reactivity series to **predict** the **outcome** of reacting metals with the oxides of other metals.

Let us look at part of the reactivity series:

MAGNESIUM
ZINC
IRON
HYDROGEN
COPPER

1) Magnesium would be able to reduce:

   Zinc oxide, iron oxide, hydrogen oxide (warm) and copper oxide

2) Iron oxide would be able to reduce **only**:

   Hydrogen oxide, copper oxide

3) Copper oxide could **not** reduce **any** of the oxides of:

   Hydrogen, iron, zinc or magnesium

## VIOLENCE OF THE REDUCTION

Let us look at reactions involving the reduction of copper oxide:

1) COPPER OXIDE + IRON ⟶ IRON OXIDE + COPPER

iron > copper in the reactivity series

So iron **takes the oxygen away** (reduces) from iron oxide.

Sometimes, we may say that iron has a greater **affinity for oxygen,** than copper.

This is why the reactivity series is sometimes called the **affinity series** – you could meet both terms.

When this reaction happens, a red glow spreads throughout the mixture and copper is clearly seen at the end.

2)

| |
|---|
| COPPER OXIDE + MAGNESIUM ⟶ MAGNESIUM OXIDE + COPPER |

Once this reaction starts it is quite violent and specks of magnesium oxide and copper are 'blown away' in the reaction. Often, very little remains in the crucible or on the ceramic paper.

These two reactions tell us that:

| |
|---|
| THE FURTHER APART IN THE REACTIVITY SERIES THE TWO METALS ARE, THE MORE VIOLENT THE REACTION WILL BE |

Think of a 'tug-of-war' contest with a team of eight on one side and a team of two on the other. The contest will be very swift and decisive!

**Placing carbon in the reactivity series**

*Carbon v magnesium*

1)

| |
|---|
| CARBON + MAGNESIUM OXIDE ⟶ NO REACTION |

So magnesium > carbon, in the reactivity series:

2)

| |
|---|
| MAGNESIUM + CARBON DIOXIDE ⟶ MAGNESIUM OXIDE + CARBON |

This reaction is usually carried out by lowering some burning magnesium into a gas jar of carbon dioxide.

The magnesium splutters and black specks of carbon and white magnesium oxide can be seen.

This confirms that magnesium > carbon in the reactivity series.

*Carbon v zinc*

| |
|---|
| CARBON + ZINC OXIDE ⟶ NO REACTION |

So zinc > carbon, in the reactivity series

(In fact, carbon **will** reduce zinc oxide, but a temperature of over

# What do you know about ...

13. What conclusions can you draw about the substances referred to as A, B, C and D? B, C and D are metals.

    (a) On heating A with dry copper oxide, water and carbon dioxide are amongst the products.
    (b) On adding B to a solution of the sulphate of C, there is no reaction.
    (c) D reacts with steam to produce an oxide and hydrogen, but will not react with cold water.

14. Magnesium burns in carbon dioxide; aluminium reduces iron oxide; carbon will reduce iron oxide; aluminium oxide will not oxidise carbon.

    (a) Place the four elements aluminium, carbon, iron and magnesium in order of increasing reactivity, explaining your reasoning.
    (b) Write word equations for the reaction mentioned.
    (c) Describe the experiments you would do to determine the place of zinc in the above series.

15. Carbon dioxide will oxidise magnesium; tin will not reduce nickel oxide; nickel oxide will oxidise carbon; tin will reduce copper oxide, steam will oxidise carbon but not nickel.

    (a) Place the elements: carbon, copper, hydrogen, magnesium, nickel and tin, in their correct places in the reactivity series, giving your reasons.

16. Magnesium reduces chromium oxide; copper displaces silver from silver chloride; lead displaces copper from copper chloride; chromium reduces iron oxide; iron reduces lead oxide.

    (a) Write down the reactivity series for the six metals (most reactive first). Explain your reasoning.

17. Three black powders are labelled A, B and C. They are known to be carbon, copper oxide, and a mixture of carbon and copper oxide (but not necessarily in this order.

    (a) Describe three simple experiments which would enable you to identify these three powders correctly.

# ... the reactivity series?

# What do you know about ...

18. Manganese is an element which is shiny and conducts electricity. It does not react with cold water, but it reacts with steam to give an inflammable gas and a dark brown residue.

    (a) Is manganese a metallic or non-metallic element?
    (b) Write a word equation to describe the reaction of manganese with steam
    (c) Draw a labelled diagram to show the reaction of manganese with steam
    (d) Do you think that manganese will react with dilute sulphuric acid? Give a reason.

19. Explain the following experimental observations. Suggest identities for the hottest substances where possible.

    (a) On heating black powdered mixture, containing metal X and compound Y, the powder changes to red-brown, but there is no change in mass.
    (b) On heating black powdered mixture 2, containing non-metal Z and Compound Y, the powder changes colour to pink and there is a decrease in mass,

20. A green powder (L) on being added to dilute sulphuric acid, fizzed out and a gas (M) was given off, leaving a blue solution (N). The gas turned limewater chalky. When a lump of zinc was added to solution (N), the colour disappeared and the zinc was coated with a pink lacquer (O).

    (a) Identify L, M, N and O. Give your reasons.
    (b) Explain why the blue colour in N disappears.

21. You are given: magnesium ribbon, silver foil, copper foil, zinc foil, together with solutions of magnesium nitrate, zinc nitrate, copper nitrate and silver nitrate. All the solutions are of the same concentration.
    (a) What is the *minimum* number of experiments you would need to do in order to investigate the reactivity of all four metals? Give reasons.
    (b) Say what you would see in each experiment and write word equations where appropriate.

# ... the reactivity series?

# What do you know about ...

22. The following metals are arranged in order of increasing reactivity:

iron
zinc
magnesium
sodium

A new metal called xorium has been discovered which reacts with magnesium oxide but not with cold water.

(a) Where would you put xorium in the reactivity series? Give a reason.
(b) Would you expect xorium to react with the following substances? Give your reasons and write word equations for any reaction which takes place.

(i) hydrochloric acid
(ii) zinc oxide
(iii) steam
(iv) potassium oxide

## ... the reactivity series?

# 8

# METALS AND MINERALS FROM ROCKS

There are about 100 elements found on earth which are all listed in the periodic table.

About **90** of these are found naturally – the other ten or so are formed as a result of nuclear reactions.

Of these 90, only a **few** elements are found as elements.

**FOR EXAMPLE**

| | |
|---|---|
| diamond – pure carbon | gold |
| copper | silver |

The majority of elements are found in parts of compounds.

**A MINERAL IS A SOLID ELEMENT OR COMPOUND FOUND IN THE GROUND**

Look at the diagram below which shows the main elements which are present in the earth's crust:

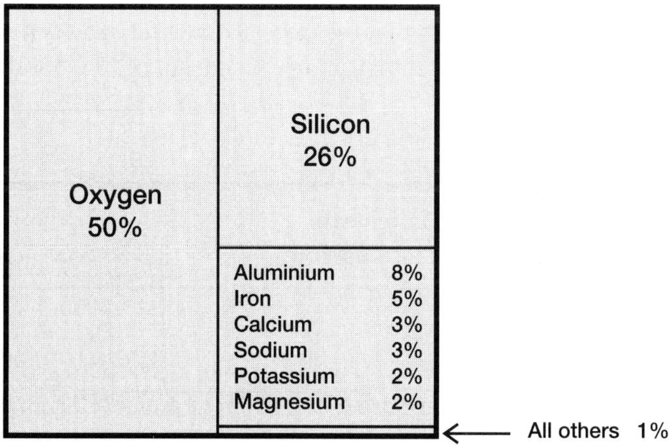


Fig. 8.1. *Composition of earth's crust*

Notice the large amounts of the elements oxygen and silicon present in the Earth's crust. In fact **silicates**, which are different forms of silicon oxides, make up most of the rocks in the earth's crust.

## A quick look at rocks

Rocks are usually **mixtures** of **minerals** and there are three main types:

### 1. SEDIMENTARY

These form as layers as there is a build-up of sediment, usually in the sea, rivers or lakes.

Small particles which have been **weathered** will be **transported** by wind/water and deposited in water in layers. Over a period of time, more layers are deposited and the increase in pressure causes the particles to be stuck together as rock.

**FOR EXAMPLE**

**Clay** forms **mudstone**.
**Sand** forms **sandstone**.
**Shells of creatures** form **limestone**.

### 2. METAMORPHIC

There are rocks which **have changed because** of the **action** of **heat** and/or **pressure**, i.e. 'squashed' and/or 'baked' without melting.

As the Earth's crust moves, so rocks can be carried deeper into the Earth's crust. This means that the once sedimentary rocks now become harder and of a different structure because of the more extreme conditions.

**FOR EXAMPLE**

**Mudstone** changes into **slate**.
**Limestone** changes into **marble**.

### 3. IGNEOUS

These are rocks which are produced when molten rock (**magma**) cools to form crystals.

Magma comes from the mantle which is below the Earth's crust and it rises through cracks and faults in the crust. Some of the magma rises and *cools very slowly* to form *large crystals* which never reach the surface. **Granite**, a very hard rock, is made this way.

Some of the magma reaches the surface, e.g. at a **volcano**, and erupts as **lava**. The lava *cools quickly* as it comes in contact with the air or sea and forms rock which consists of *small crystals*. Most of the ocean floors are made of this type of rock (known as **basalt**).

Rocks are forever changing. They are weathered at the surface of the Earth, are deposited as sediments, changed as they move deeper into the Earth's crust. Some rocks will become magma, be erupted as lava and so there is a continual **ROCK CYCLE**. (This is discussed in more detail in the Physics & Earth Science book.)

## Ores

As we have seen, a few unreactive metals (gold, silver, copper) can be found as metals – i.e. found **native**.

Most metals are found as a mixture of rock and metal compound (mineral) which we call an **ore**.

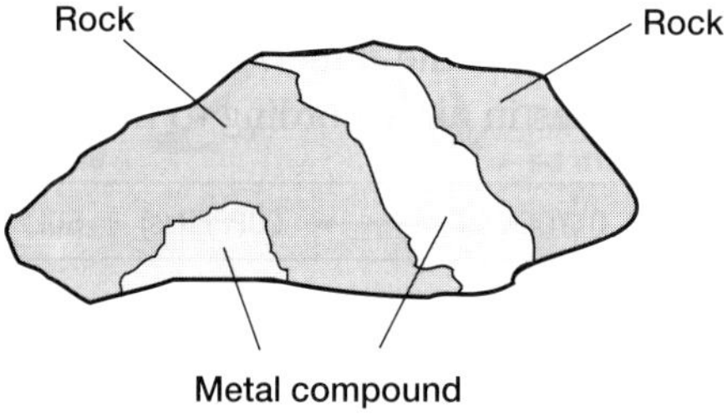


Fig. 8.2. *Diagram of ore*

## Extraction of metals

The method of extraction will depend upon the *reactivity* of the metal.

| THE HIGHER A METAL IS IN THE REACTIVITY SERIES, THE HARDER IT IS TO EXTRACT IT FROM ITS ORE |
|---|

Look at Table 8.1 on the following page which shows the metals listed in order of reactivity.

The electric current causes the copper sulphate to split up into copper ions and sulphate ions. The copper ions move to the cathode and are deposited. The sulphate ions move to the anode and 'collect' more copper from the anode. This 'collected' copper becomes copper ions and so moves to the cathode. The process continues until all of the copper has been removed from the anode. As *only* copper ions are moved to the cathode, any impurities (i.e. non-copper substances) merely fall to the bottom as a sludge. This sludge is collected and processed separately as indeed, some of the 'impurities' could well be gold or silver.

The cathodes are removed from the electrolytic cell and the sheets of 99.99% copper are stripped off the electrode (often sheets of titanium), washed – as the electrolyte is acidified copper sulphate solution – and then it is ready for melting down into convenient shapes ready for the manufacture of various items.

**Copper from malachite**

The mineral malachite contains copper carbonate and it is often considered that historically this is how copper was extracted in early times. It would have been made in early fires where wood was the fuel so the mineral would have been heated up surrounded by **carbon** in the form of charcoal.

Although the ancient smelters would have been unaware of the details, the extraction would have been in two stages.

*Stage I*

The heat from the fire would have decomposed the malachite (copper carbonate).

$$\text{COPPER CARBONATE (s)} \xrightarrow{\text{HEATING}} \text{COPPER OXIDE (s)} + \text{CARBON DIOXIDE (g)}$$

*Stage II*

The copper oxide would be heated in the fire with carbon (charcoal). The copper formed would be molten and run to the bottom of the fire and be collected from the ashes when they had become cool after the fire had died down.

$$\text{COPPER OXIDE (s)} + \text{CARBON (s)} \longrightarrow \text{COPPER (s)} + \text{CARBON DIOXIDE (g)}$$

In ancient times, copper would have limited uses as it is a relatively soft metal. It would not have been hard enough for axes or tools. However, it is thought that the ancient Egyptians used 'mirrors' of polished copper to provide light for the workers and artists working deep underground constructing and decorating the elaborate tombs of the Pharaohs.

When, however, molten copper was mixed with molten tin, the **ALLOY BRONZE** was formed. This is a hard material and the **BRONZE AGE** refers to that period in history when weapons and tools were formed from this alloy.

### TIN

Tin, like copper, is low in the reactivity series and is easily extracted using the woodsman's charcoal fire.

It is formed as the mineral **CASSITERITE** which is tin oxide.

| | | | | |
|---|---|---|---|---|
| TIN OXIDE (s) + | CARBON (s) | ⟶ | TIN (s) + | CARBON DIOXIDE |
| (Cassiterite) | (Charcoal) | | | |

### LEAD

As with copper and tin, lead, being low in the reactivity series, is relatively easy to extract from its ore.

The chief source of lead, is the mineral **GALENA** which is lead sulphide.

Like **malachite**, the sulphide needs to be changed into the oxide before the reduction can take place.

Think of the air conditions around the location of lead smelting. Although lead could be extracted by the woodsman's charcoal fire – this would not be a fire to gather and sing songs around!

#### Stage I

| | | | | |
|---|---|---|---|---|
| LEAD SULPHIDE (s) + | OXYGEN (g) | ⟶ | LEAD OXIDE (s) + | SULPHUR DIOXIDE (g) |
| (Galena) | (Air) | | | |

#### Stage II

LEAD OXIDE (s) + CARBON (s) ⟶ LEAD (s) + CARBON DIOXIDE (g)

Note the similarities in the extraction of copper, lead and tin.

1. If the metal ore is a carbonate or sulphide – heat it in air to change the mineral to an **oxide** (or metal in the case of copper sulphide).
2. The reduction:

| |
|---|
| METAL OXIDE + REDUCING AGENT ⟶ METAL + WASTE GASES |

In all these cases, the reducing agent would have been carbon which was easily obtained from the surrounding forests. Remember that forests in early times were much more extensive than they are today.

If your knowledge of history is shaky, then chemistry will help you to remember that the **BRONZE AGE** came *before* the **IRON AGE**.

Bronze (the alloy of copper and tin) was obtained at the temperatures which would have been easily achieved in the charcoal fires of the smelters.

Iron, however, needs a greater temperature to extract it from its ore. This means that a higher level of technology was needed, in terms of specially designed furnaces.

## EXTRACTION OF IRON

Early iron works were located in areas where there was a good supply of iron ore and carbon (from charcoal). Thus in England, sites of ancient ironworks can be found in ancient forests which grew where there was a supply of iron ore.

The main ore used in the extraction of iron is HAEMATITE (iron oxide).

| |
|---|
| IRON OXIDE (s) + CARBON (s) ⟶ IRON (s) + CARBON DIOXIDE (g) |

Later and to this day, coke has replaced charcoal as the source of carbon. This is obtained from coal.

**Limestone** is also used:

(a) as a source of carbon dioxide
(b) to combine with the excess rock and hence separate it from the iron. This combination is called **SLAG**, which being less dense than iron floats on top of the molten iron in the furnace. This can be run away separately and used as a **road-building** material.

The extraction of iron is carried out in a **BLAST FURNACE**, the blast being hot air (containing oxygen).

**Raw materials**

A mixture of crushed:

| | | |
|---|---|---|
| 1) | iron ore | called the **CHARGE** |
| 2) | limestone | |
| 3) | coke | |
| 4) | a blast of hot air – oxygen | |

Although carbon will reduce iron oxide if the temperature is high enough, the reducing agent used is **CARBON MONOXIDE** so there are *three main* reactions which take place in the blast furnace:

1) making the reducing agent (carbon monoxide)
2) reduction of iron oxide
3) formation of slag

**1. Making the reducing agent (carbon monoxide)**

(a) Coke burns in the hot blast to form carbon dioxide.

CARBON (s) + OXYGEN (g) ⟶ CARBON DIOXIDE (g)

(b) Limestone (calcium carbonate) decomposes when heated to release carbon dioxide.

CALCIUM CARBONATE (s) —HEATING⟶ CALCIUM OXIDE (s) + CARBON DIOXIDE (g)

(c) The carbon dioxide passes over more hot coke to form carbon monoxide.

CARBON DIOXIDE (g) + CARBON (s) ⟶ CARBON MONOXIDE (g)

**2. Reduction of the iron oxide**

Carbon monoxide reduces iron oxide and the iron formed is molten at the high temperature (1800°C) produced, and runs to the bottom of the furnace where it collects until it is **tapped** (run off).

IRON OXIDE (s) + CARBON MONOXIDE (s) ⟶ IRON (l) + CARBON DIOXIDE (g)

### 3. Formation of slag

The 'rocky' part of the haematite will be mainly composed of sandy substances such as silicon oxide. This will combine with the calcium oxide (from 1), to form slag.

| | | | |
|---|---|---|---|
| SANDY ROCK + | CALCIUM OXIDE | ⟶ | SLAG |
| (Silicon oxide) | | | (Calcium silicate) |

Again, slag will be molten at the high temperature and so drips to the bottom of the furnace where it floats on the more dense iron.

The iron from the blast furnace is very brittle as it is impure. The main impurity is carbon (about 4%), although there are often traces of phosphorous and sulphur.

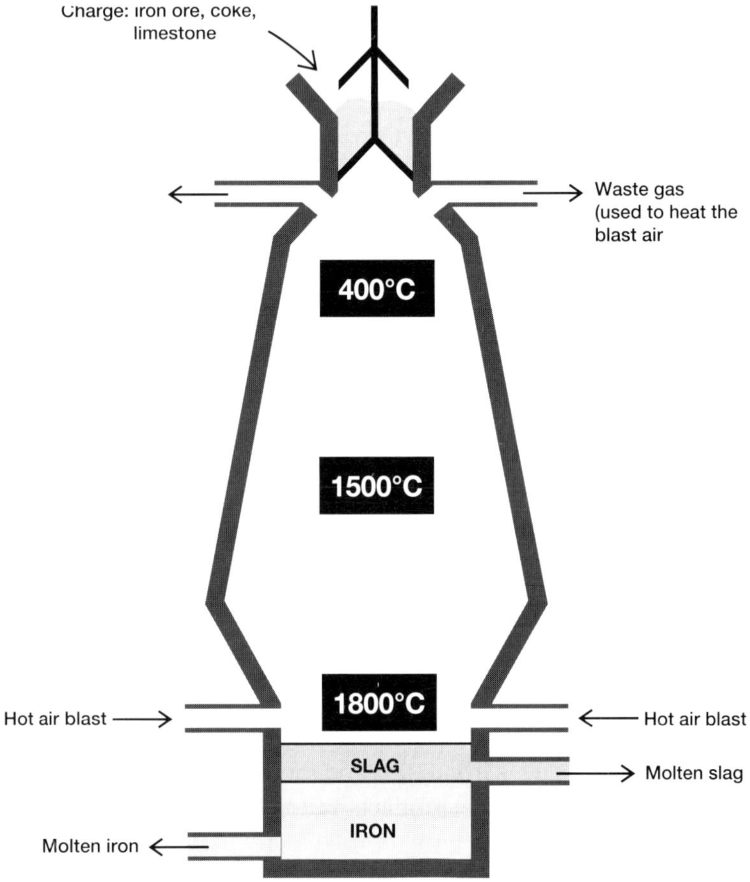


Fig. 8.5 *A blast furnace*

This impure iron (often called **CAST IRON**) does have limited uses and it can be melted down and poured into moulds for use as ornamental railings and drain covers.

Most of the cast iron produced in the blast furnace will be **converted** into **steel**.

Steel is mainly iron, but it has had most of the carbon – and certainly other non-metal impurities – removed from it. It is generally mixed with other metals to form alloys

**FOR EXAMPLE**

IRON + CHROMIUM RESULTS IN THE ALLOY **STAINLESS STEEL**

The formation of steel from iron is done in a **CONVERTER**.

Molten iron is poured into a converter and oxygen is blown into it through an **oxygen lance**.

The impurities of carbon and other non-metal impurities burn fiercely at the surface of the liquid. When these have finished burning, the oxygen lance is removed and small amounts of other metals are added to make the alloys.

## EXTRACTION OF ZINC

The main source of zinc is from the mineral **SPHALERITE** (zinc sulphide). This, being more reactive than iron, will be reduced using carbon monoxide.

You should by now realise that the extraction will be in two main stages:

1) changing the sulphide into the oxide;
2) reduction of the oxide using carbon monoxide.

### Uses of zinc

1) For galvanising iron (see section on rusting)
2) For mixing with copper to form the alloy, **brass**.

## More chemicals from rocks

As well as extracting metals from rocks, we mine (or quarry) a great variety of rocks which contain minerals from which we extract many important chemicals. A few are listed below.

| MINERAL | MAIN CHEMICAL COMPOUND |
|---|---|
| Limestone, calcite<br>Marble, chalk<br>Iceland spar | Calcium carbonate |
| Gypsum | Calcium sulphate |
| Rock salt, halite | Sodium chloride |
| Quartz | Silicon dioxide |

### LIMESTONE, CHALK, MARBLE

These are all forms of **calcium carbonate**. **Limestone** and **chalk** are **sedimentary rocks** which have been made by the deposition of the shells of creatures which lived in the ancient seas. Close inspection of these minerals with a hand lens will sometimes reveal fossils of these ancient creatures – if you are lucky.

Marble will have *started* as a sedimentary rock in the same way as limestone, but it will have been moved deeper into the Earth's crust where increased heat and pressure will have changed it into the much harder **metamorphic rock**, **marble**.

## Uses of limestone

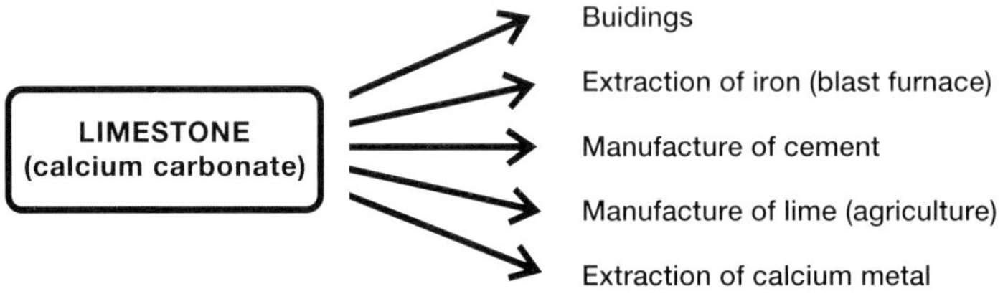


Fig. 8.6

### BUILDING

Some forms of limestone are hard enough to be used as 'bricks' for building. In England, much of this hard limestone comes from the Purbeck and Portland area in Dorset. Whilst the buildings have an attractive, clean white look, the one disadvantage of using limestone is that being calcium carbonate, limestone is prone to attack from acid rain.

## EXTRACTION OF IRON

We have just seen that limestone is one of the ingredients which make up the charge in blast furnaces. As the world's demand for steel (and hence iron) increases, so a vast tonnage of limestone will be mined for this purpose.

## MANUFACTURE OF CEMENT

Over 1000 million tonnes of cement are made in the world every year and this material is the basic ingredient in two important substances used in the construction industry – **mortar** and **concrete**.

### 1. Mortar – a mixture of cement, sand and water

Mortar is used to 'bind' bricks, stones or blocks together in all forms of buildings. On your house, the 'lines' between the bricks will consist of mortar and collectively, they are known as the **pointing**.

### 2. Concrete – a mixture of cement, sand, small stones and water

The technical name for the small stones is **aggregate** and these will be added to give 'body' to the concrete.

Concrete is one of the major construction materials in the world and you can see vast quantities all around you – from your concrete path, to buildings, bridges, elevated motorways and dams. The list is endless.

Concrete is very strong and extremely resistant to compression. This is why it is such a good material for foundations of construction, as it can withstand the constant weight.

Even though it is strong, it is not so good at withstanding lateral (sideways) forces (pushes and pulls). This problem is overcome by the insertion of a network of steel rods inside the concrete to produce **reinforced concrete**. It will be this form of concrete which is used in elevated motorway construction.

Cement is made by roasting a mixture of **limestone** and **clay** in a **rotating lime kiln**. A small amount of gypsum (calcium sulphate) will be added to make the cement easier to work with. It slows down the 'setting' process.

## Gypsum (calcium sulphate)

This is generally found as crystals which are gently heated to lose most of the water of crystallisation. The product left is a powder which forms the basis of plaster. The smooth coverings on walls will be plaster.

When plaster is added to water it warms up and as it dries it sets hard.

## A bit of a lather – hardness of water

If you live in an area where there are limestone rocks, then the chances are that you have **hard water**.

| |
|---|
| HARD WATER IS WATER WHICH CONTAINS DISSOLVED CALCIUM (OR MAGNESIUM) COMPOUNDS |

A lather is the soapy, bubbly foam which is produced when you agitate soap in water.

| |
|---|
| SOFT WATER – EASY TO PRODUCE A SOAPY LATHER<br>HARD WATER – HARD TO PRODUCE A SOAPY LATHER |

### TEMPORARY HARDNESS – CAN BE REMOVED BY BOILING

Temporary hardness is caused when water contains **calcium hydrogen carbonate**. We have come across this compound before. It is the soluble salt which is formed when carbon dioxide is bubbled into chalky lime water, which cleans.

Rain is a weak solution of carbonic acid and this will react with limestone to form calcium hydrogen carbonate.

| | | | | |
|---|---|---|---|---|
| CARBONIC ACID | + | CALCIUM CARBONATE | ⟶ | CALCIUM HYDROGEN CARBONATE |
| (Rain) | | (Limestone) | | (Hardness in water) |

When temporary hard water is boiled, carbon dioxide is given off and calcium carbonate re-forms as the 'fur' in kettles and pipes.

### PERMANENT HARDNESS – NOT REMOVED BY BOILING

This is caused by **calcium sulphate** (or magnesium sulphate) being dissolved in the water.

Water which is permanently hard is passed through a **water softener** which removes the calcium ions and replaces them with sodium ions. This is known as **ION EXCHANGE**.

Sodium salts do not cause hardness of water.

## Rock salt (sodium chloride)

You may well have come across crushed rock salt which is spread over roads during the winter to prevent them from being icy.

You will certainly have purified rock salt and obtained crystals of salt similar to the salt which you put on your chips.

Sodium chloride is however an important raw material which is used by the chemical industry on a vast scale. Some of its uses are listed below:

### 1. EXTRACTION OF SODIUM AND CHLORINE

Molten sodium chloride is electrolysed with sodium metal and chlorine gas being the products.

Chlorine is used not only as a water purifier, but also in the manufacture of plastics (PVC) and dry cleaning fluids.

### 2. PRODUCTION OF SODIUM CARBONATE

As we saw (in the section on limestone), this is an important material which is produced using sodium chloride as one of its ingredients.

### 3. PRODUCTION OF SODIUM HYDROXIDE (CAUSTIC SODA)

Sodium chloride will be dissolved in water to form **brine** and this will be electrolysed to form sodium hydroxide, chlorine and hydrogen.

Over 26 million tonnes of sodium hydroxide are used every year.

Sodium hydroxide is an important industrial cleaner and is also used in the manufacture of detergents, in the manufacture of textiles and as a 'feedstock' for making other chemicals.

## Sand and quartz

These are both different forms of silicon oxide. **Sand – silica or silicon oxide** – is one of the most abundant minerals on the Earth. It is used in the building and construction industry (to make mortar and concrete) and you may have used paper covered with grains of sand (sandpaper!) to smooth wood.

A huge use of sand is in the manufacture of glass.

## QUARTZ – SILICON DIOXIDE

This is a very special form of crystal which is treated in such a way that it will give **regular pulses** when a *small* current is passed through it. These crystals are the heart of quartz watches which are so abundant these days.

# What do you know about ...

1. (a) What is a mineral?
   (b) Give an example of a mineral and say why it is worth mining.

2. Name the three main types of rock. Give an example of each and say how each type is formed.

3. (a) Which type of rock are you likely to find fossils in? Give a reason.
   (b) Which type of rock *never* contains fossils? Give a reason.

4. Limestone and marble contain the same chemical compound but came from different types of rock.

   (a) What is the name of the chemical compound found in both types?
   (b) What are the different types of rock limestone and marble and give an explanation for the difference.

5. (a) What is an ore?
   (b) Some ores are said to be low grade. What does this mean?

6. (a) Which are the two most common elements in the Earth's crust?
   (b) Which is the most common metal?
   (c) Some metals can be found native. What does this mean? Give an example.

7. How would you obtain some copper metal from malachite (copper carbonate)? Say:

   (a) what materials you would use
   (b) how you would extract the metal;
   (c) give word equations for any reactions which take place.

8. Iron is higher in the reactivity series than lead. The chief ore for the extraction of lead is galena (lead sulphide) and lead oxide is reduced by carbon.

   (a) How do you change lead sulphide into lead oxide? Write a word equation to illustrate your answer.
   (b) What is the likely source of carbon?
   (c) Write a word equation for the reduction of lead oxide.
   (d) Would iron reduce lead oxide? Give a reason.
   (e) Explain why iron is not used to produce lead.

# ... metals, minerals and rocks?

# What do you know about …

9. Explain why a knowledge of chemistry will help you to remember that the bronze age came before the iron age.

10. Iron is made in a blast furnace. Materials loaded into the furnace are called the charge. Molten liquids are 'tapped' from the bottom of the furnace.

    (a) Why is the furnace called a blast furnace? What is the purpose of this special feature?
    (b) What materials form the charge and what is the purpose of each material?
    (c) What waste gases come out of the furnace?
    (d) (i) What is the main reducing agent in the process?
        (ii) How is this formed?
        (iii) Write a word equation for the making of the reducing agent.
    (e) Write a word equation for the main reduction in the furnace.
    (f) What two liquids are tapped at the bottom of the furnace?

11. The iron produced in a blast furnace contains carbon, sulphur and other non-metal impurities.
    (a) What effect do these impurities have on the properties of this type of iron?
    (b) What is this type of iron called?
    (c) Name two uses for this type of iron.

12. (a) What is steel?
    (b) How is it formed? Do not include processes involving basic rocks in your answer.
    (c) Name two different types of steel, say what each type is used for and say why each type is suitable for the use which you have selected.

13. Aluminium, copper, glass, iron, lead, plastic, rubber.

    From this list of materials, choose one that would be most suitable for the uses given below. In each case, give a reason for your choice.

    (a) a picnic mug
    (b) the inner wires of an electric cable
    (c) the guttering around the roof of a house

## … metals, minerals and rocks?

# What do you know about ...

(d) a bucket to carry red hot ashes
(e) a girder to support a bridge
(f) wings of an aeroplane.

14. The chemical symbol for lead is Pb (Plumbum) – hence the word plumber for someone who fixes and mends pipes.

Copper pipes have now largely replaced lead pipes for household water.

Give one advantage and one disadvantage of using lead for water pipes and say why copper is a better alternative.

15. Below is a table which tells you about the extraction of three metals.

| Metal | Extraction |
|---|---|
| Aluminium | From aluminium oxide by a chemical reaction involving large amounts of electricity |
| Gold | By crushing the ore and washing it |
| Iron | From iron oxide by a chemical reaction with carbon |

(a) Put the three metals in order of (increasing) reactivity, giving your reasons.
(b) Aluminium is more abundant than iron, but aluminium is more than ten times as expensive as iron. Why do you think that this is so?
(c) Why is gold much more expensive than either iron or aluminium?

16. Early this century, iron saucepans were used for cooking. Nowadays people often use saucepans made of aluminium.

(a) Aluminium cannot be obtained from its oxide by heating it with coke. Why not?
(b) Give one advantage of using aluminium rather than iron for making saucepans.
(c) You have two saucepans which are the same size. One is made of aluminium, whilst the other is made of iron. The iron one is the more expensive of the two.

Why do you think the iron one is more expensive and why is this fact surprising?

# ... metals, minerals and rocks?

# What do you know about ...

21. Some soluble calcium compounds are responsible for the temporary hardness of water. Limestone (calcium carbonate) is insoluble in water. Rain always contains small amounts of dissolved carbon dioxide.

(a) What is meant by hardness of water?
(b) Why can we refer to a hardness of water as being temporary?
(c) Water which flows over limestone rocks may well exhibit temporary hardness. How can this be, given that limestone is insoluble in water?
(d) State one problem that temporary hardness of water causes and say how the problem arises.
(e) If you are an avid tea drinker, would you prefer your water to be permanently hard or temporarily hard? Give a reason for your answer.

## ... metals, minerals and rocks?

# Further questions on ...

1. The mineral malachite contains three elements.
   (a) State what they are.
   (b) Describe and explain experiments which you could do, in order to show the presence of each of these elements in malachite.

2. Some metals occur naturally, but most occur as compounds requiring a variety of methods of extraction.
   Below are approximate discovery dates of some metals.

| | |
|---|---|
| gold, silver | early civilisation |
| copper, tin | bronze age |
| iron | iron age |
| zinc | AD 1 |
| potassium, sodium | 1807 |
| aluminium | 1825 |

   The hot furnace was developed about 5 BC; sources of electric current about 1800.
   Using this information, explain the ways by which the metals were discovered.

3. Explain the facts in the following two sentences.

   Whilst gold mines are well known, gold chloride mines have not been discovered. Salt mines are easily found, but sodium mines have not been discovered.

4. Copper, silver and gold are sometimes called the 'coinage' metals.
   (a) Suggest why these particular metals were often used to make coins.
   (b) Why isn't our present £1 coin made of gold?
   (c) Iron is one of our cheapest metals. Why isn't it used to make our penny coins?

## ... metals, minerals and rocks?

# Further questions on ...

5. Pure (99.99%) copper can be obtained from the mineral chalcocite (copper sulphide). 2.5% of the rock mined is chalcocite.
   (a) How many tonnes of rock have to be mined to obtain 1 tonne of chalcocite?
   (b) The following three words are often used to describe the obtaining of pure copper:

   electrolysis, decomposition, reduction

   Explain each word in terms of the reactions going on. You should use word equations where possible.

6. Limestone is a sedimentary rock which largely consists of calcium carbonate.
   (a) Write a word equation for the conversion of limestone to quicklime.
   (b) Quicklime can also be converted into slaked lime. Suggest how this might be done and write a word equation for the process.
   (c) Is the process which you have mentioned in (b) a chemical or a physical change? Explain your reasons detailing any evidence you would seek to obtain.
   (d) Write down a large scale use of:
       (i) limestone
       (ii) quicklime
       (iii) slaked lime
   (e) (i) What is meant by the term 'sedimentary work'?
       (ii) How is limestone formed?

7. An alloy is a mixture of two or more metals. The diagrams below show the difference in structure between a pure metal and an alloy.

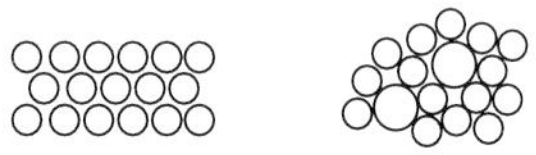

   (a) What do the circles of different size represent?
   (b) An alloy is generally stronger than a pure metal when it is pulled. Using the information in the diagrams above, explain why this is so.
   (c) Give examples of two other properties which would be affected by the different structures of an alloy and a pure metal and state how properties would be affected.
   (d) Give an example of two alloys and in each case, say what materials go to make up the mixture.

# ... metals, minerals and rocks?

# Further questions on ...

8. Mortar is made by mixing 1 part of cement (a calcium oxide and clay mixture) with three parts of sand into a paste with water.

The temperature rises during the mixing. When applied to the bricks, the mortar first sets and then slowly reacts with carbon dioxide in the atmosphere to form calcium carbonate which effectively binds the grains of sand.

(a) Why does the mortar rise in temperature as it is being mixed? Use a word equation to support your answer.

(b) Two samples of mortar are taken from the inside of a building and shown to have been used at quite different times.

   (i) Describe how you might have arrived at this conclusion.

   (ii) Why would your method not work on a very old building?

   (iii) Why does mortar on the outside of a building have to be replaced more often than the mortar on the inside of the same building?

9. Describe with clear diagrams and numbered steps an extraction that you have carried out in the laboratory to obtain a metal from a rock.

Your answer must contain all the following words in such a context that their meaning is clear.

ore
refine
oxidise
element
compound
combination
reaction

## ... metals, minerals and rocks?

# ELECTRICITY IN CHEMISTRY — AN INTRODUCTION

9

We saw (in chapter 8) that electricity is used to extract reactive metals, e.g. sodium, potassium and calcium from their ores and that, despite the fact that aluminium compounds are the most abundant on the Earth's surface, the large scale extraction of this important metal was only possible when large amounts of electricity were available.

You will also know that your torch lights up because the chemicals in an electric cell (or battery of cells) react to produce an electric current.

These are but two examples of the action of electricity in chemistry. There are more, for most chemical reactions involve electric charge in one way or another and this is why it is a good idea to have a simple appreciation of the importance of electricity in chemistry at this stage.

## Electric charge and the atom

We saw (in chapter 1) that all substances are made from one or more **elements**; that there are about 100 different elements and that the smallest particle of an element is an **atom** of that element. We also saw that atoms were made from three different types of sub-atomic particle:

**protons** – carrying a positive charge
**electrons** – carrying a negative charge
**neutrons** – same mass as a proton but carrying no charge at all

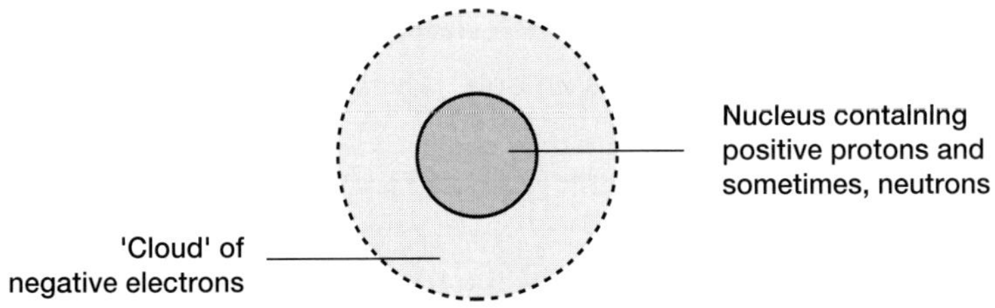


Fig. 9.1 *General diagram of an atom*

The electrons are arranged in layers or **SHELLS** around the nucleus.

There are many features of the arrangements of electrons in their shells which you will study in more detail as you progress through your chemistry studies at GCSE level and beyond. However, at this stage, it is sufficient to be aware of the following:

1. There is a maximum number of electrons that can be found in any particular shell, e.g. the maxim number in the shell nearest to the nucleus is 2.

2. The number of electrons in the *outer* shell **is the same** as the element's group number in the periodic table.

We have concentrated on the electrons because it is these which cause substances to bind together and/or react – particularly the **outer** electrons of atoms.

Let us now look at how the outer electrons of certain elements enable atoms to bind together.

## Formation of ions

IN **ATOMS**: THE NUMBER OF **PROTONS** (+VE) IS **EQUAL** TO THE NUMBER OF **ELECTRONS** (–VE)

IN **IONS**: THE NUMBER OF **PROTONS** (+VE) **DOES NOT EQUAL** THE NUMBER OF **ELECTRONS** (–VE)

### METAL IONS ARE POSITIVE

This is because metal atoms **lose** electron/s from their outer shells and so the number of protons is **greater** than the number of electrons.

**FOR EXAMPLE**

Sodium (Na) atom loses an outer electron to form a Sodium Ion ($Na^+$).

## Why metals behave the way they do

**REMEMBER**

All metals:
1) conduct heat
2) conduct electricity
3) are flexible

They are able to have the properties listed above, because of their structure.

In metals, the positive ions are situated in a regular framework (called a lattice). The discarded electrons are free to roam (see below).

## Sodium metal

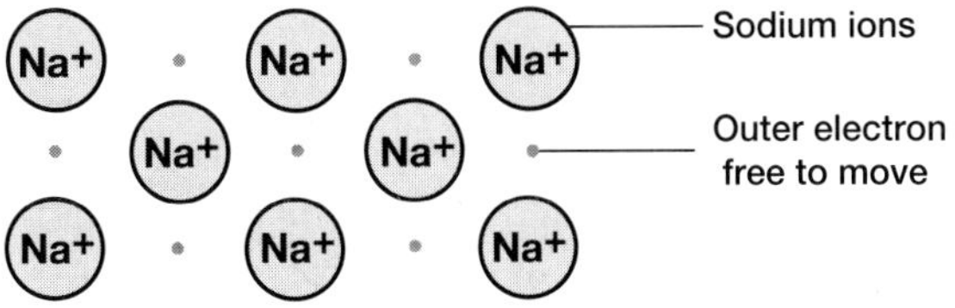


Looking at the properties of sodium listed above:

### 1. CONDUCT HEAT

'Free' electrons are able to move more rapidly as heat moves through the metal (i.e. is **conducted**).

### 2. CONDUCT ELECTRICITY

The 'free' electrons move through the metal, each one carrying a negative charge. When they all move in the **same direction**, then an **electric current** is flowing through the metal.

### 3. FLEXIBILITY

The layers of ions are able to 'slide' past each other whilst the 'sea of free electrons':

(i) acts as a lubricant
(ii) holds the lattice together

## Non-metal ions are negative

This is because when non-metal atoms react with metal atoms, there is a **transfer** of electron/s **from** the metal atoms **to** the non-metal atoms. In effect, this makes:

1) A metal atom **lose** an electron and so become a **positive metal ion**;

2) A non-metal atom **gain** an electron and so become a **negative non-metal** ion.

This can be seen clearly by looking at the reaction between sodium and chlorine to form the compound, sodium chloride.

A SODIUM ATOM LOSES AN ELECTRON TO FORM A SODIUM ION

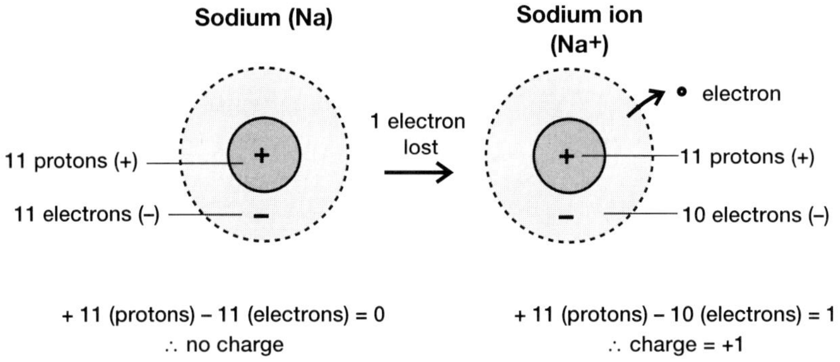


Fig. 9.2

This is usually expressed as:

(i) Word equation: SODIUM ATOM ⟶ SODIUM ION + ELECTRON

(ii) Using formulae: $Na \longrightarrow Na^+ + e^-$

A chlorine atom gains an electron (from the sodium atom) to form a chlorine ion.

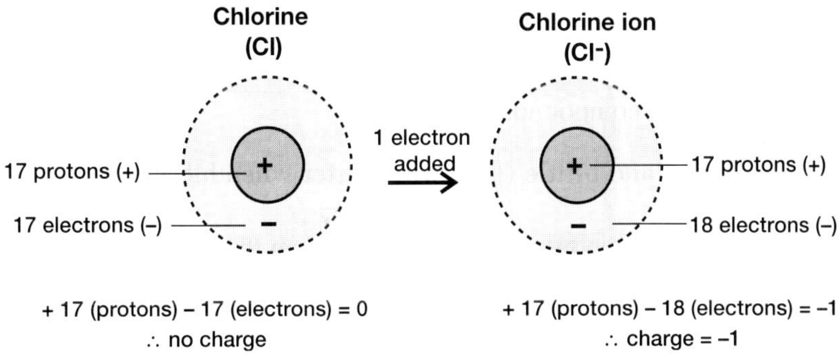


Fig. 9.3

This is expressed as:

(i) Word equation: CHLORINE ATOM + ELECTRON ⟶ CHLORINE ION

(ii) Using formulae $Cl + e^- \longrightarrow Cl^-$

## THUS

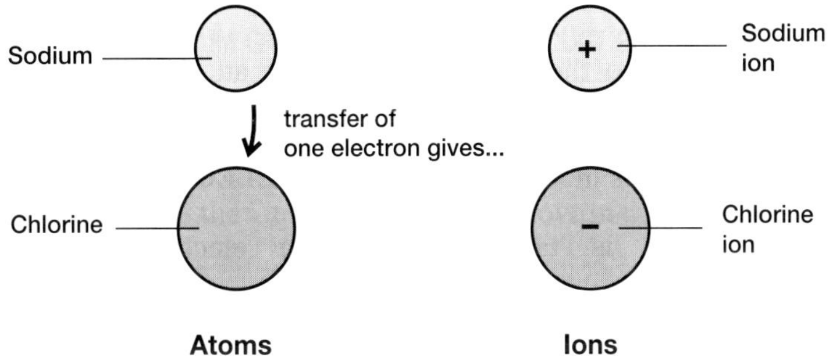


Fig. 9.4

The ions, being oppositely charged, attract each other electrostatically and form a lattice of sodium chloride.

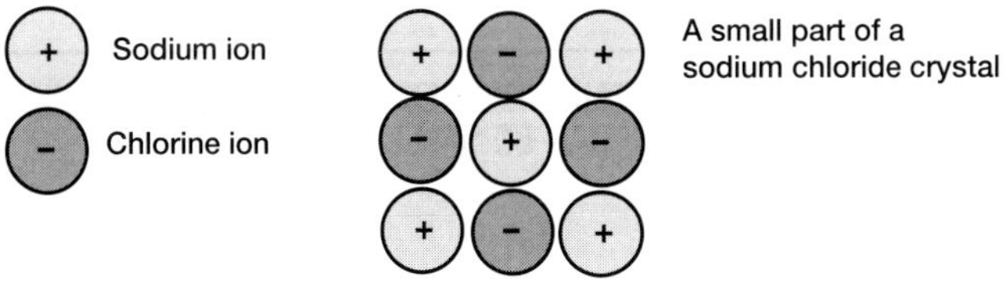


Fig. 9.5

There are problems with this cell.

1. Hot chlorine and sodium must be kept apart – otherwise they will react together to reform sodium chloride.
2. Sodium boils at 883°C – this means that care has to be taken to stop the molten sodium from evaporating.

Two important measures are taken when this reaction is carried out on an industrial scale:

1. The chlorine is led away from the cell as soon as it is produced at the anode.
2. Calcium chloride is added to the sodium chloride to drop its melting point to around 600°C.

## ELECTROLYSIS OF SODIUM CHLORIDE SOLUTION

On the face of it, this would seem a simple reaction as positive sodium ions ($Na^+$) would move towards the negative electrode (cathode) and the negative chloride ions ($Cl^-$) would move towards the positive electrode (anode).

In fact, this is what initially happens.

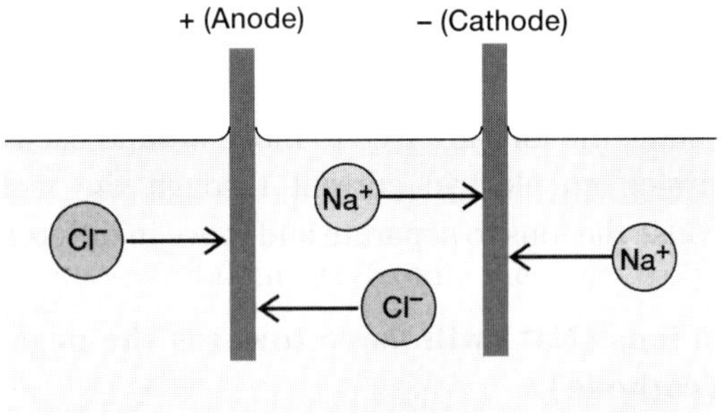


Fig 9.8

**RESULT**

**At anode** – chlorine gas is produced.

**At cathode** – sodium metal should be produced.

However, as sodium is more reactive than hydrogen, sodium ions are more difficult to displace than the hydrogen ions which go to make up water.

So **HYDROGEN** is produced here and the sodium ions combine with the remaining part of the water molecule to form sodium hydroxide.

## USING ELECTROLYSIS TO PURIFY METALS

### Purifying copper

Copper is produced by reducing copper oxide to copper not pure enough to be used for electrical wiring.

It is purified using electrolysis.

The **impure copper** is made into slabs and is used as the **anode**.

Sheets of titanium are often used as the cathode. The electrolyte used is acidified copper sulphate solution.

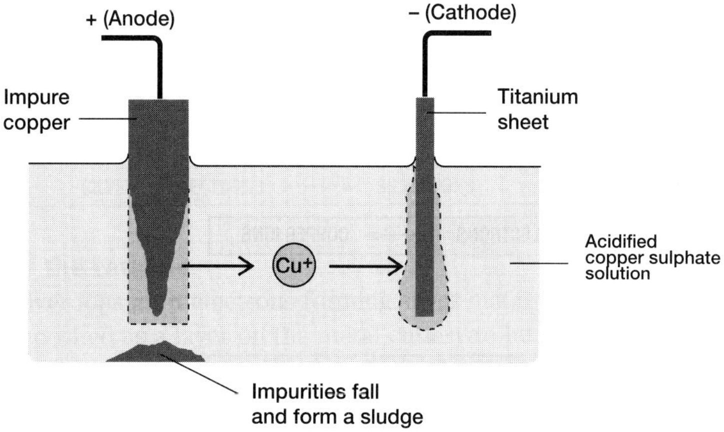


Fig. 9.9

Pure copper forms as a deposit on the titanium cathodes.

After working to remove any traces of acidified copper sulphate solution the pure copper is stripped off the titanium sheets, and melted into ingots ready to be made into wires at a later date.

Copper produced by this method is 99.99% pure and is suitable for electrical purposes.

# Electroplating

## COPPER PLATING

If you wished to copper-plate a metal paper clip, then you would use a set-up similar to that used in the purification of copper (see Fig. 9.9).

## Electropainting

This was initially developed to enable the **inside** of vehicle petrol tanks to be painted. Before this development, petrol tanks were made in two halves, each half being painted before the two halves were welded together.

The petrol tank is made the cathode and the paint particles are positively charged by the anode.

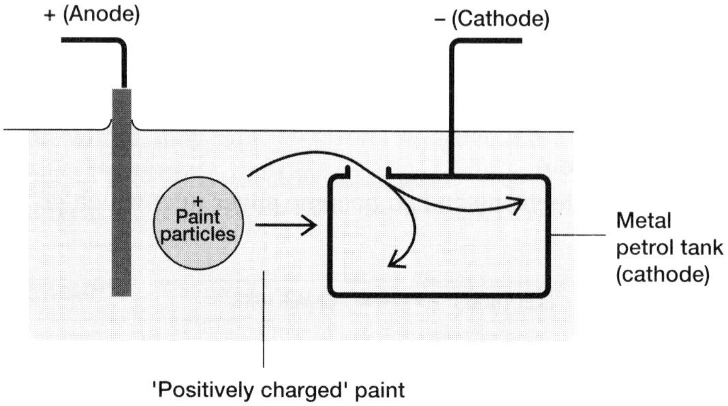


Fig. 9.12

## Chemical reactions producing an electric current

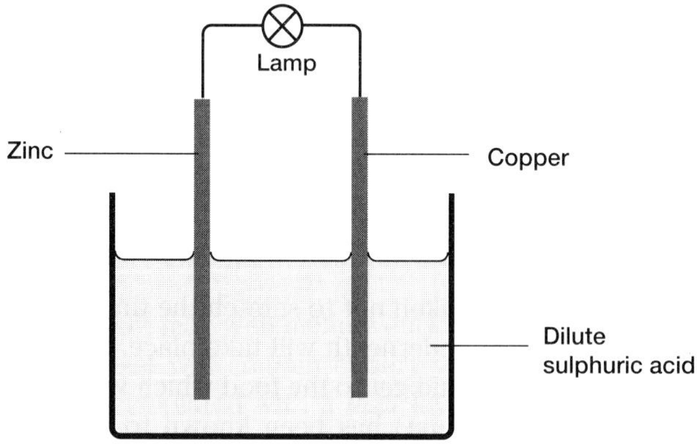


Fig. 9.13

The above arrangement is known as a **cell**. If you connect two/more cells together, you will make a **battery**.

Zinc is more reactive than copper and will form ions in solution more easily.

When these two metals are dipped into dilute sulphuric acid, it is the zinc which begins to dissolve and form ions.

ZINC ⟶ ZINC IONS + ELECTRONS

The 'spare' electrons from the reaction travel up the zinc rod and through the connecting wire as an electric current. Fig. 9.13 shows that it is possible to light a small lamp this way.

The electrons will flow through the wire, because a **POTENTIAL DIFFERENCE** exists between the two metal rods when they are both dipped in acid.

Potential difference is measured in volts.

Different metals will give different potential differences and these can be measured. Some results are given below.

The order of metals in Table 9.1 should be familiar to you as they are the same as the reactivity (or affinity) series.

In fact, this series is called the **ELECTROCHEMICAL SERIES**.

**TABLE 9.1**

| Metal pairings | Potential difference (volts) |
|---|---|
| Copper–magnesium | 1.2 V |
| Copper–zinc | 0.7 V |
| Copper–iron | 0.3 V |
| Copper–copper* | |

* Note that when both metals are the same, there is no potential difference so no current flows.

## THE DRY CELL

This is most commonly known as a battery, but this is strictly incorrect when dealing with single cells which have a value of 1.5 V.

In a 'dry' cell, the electrodes are usually zinc and carbon. The carbon is the positive electrode and the outer zinc casing is the negative electrode.

The electrolyte is in the form of a paste and this is why the cell is called 'dry'.

As the reaction proceeds, the outer zinc casing wears away. This is why old 'batteries' sometimes leak and why they should always be removed from whatever they are supplying electricity to when:

1. the cells run down;
2. the electrical equipment is not going to be in use for some time.

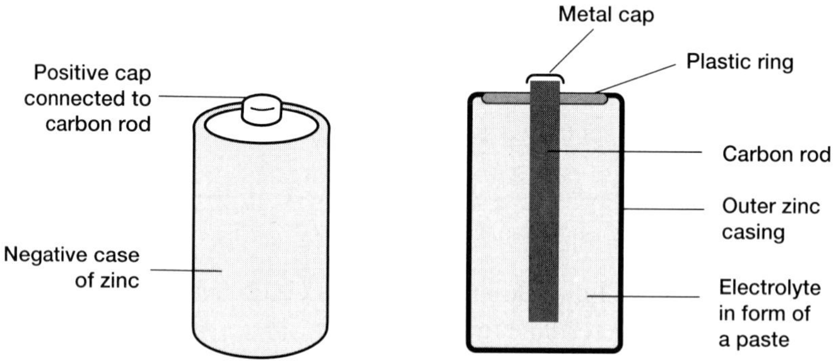


Fig. 9.14

# What do you know about ...

1. A hydrogen atom is made up of 1 proton and 1 electron.

   (a) Draw a diagram of a hydrogen atom to show this information.
   (b) What, if any, electric charge does the electron carry?
   (c) What, if any, electric charge does the proton carry?
   (d) Show, by means of labelled diagram, how a hydrogen atom can be changed into a hydrogen ion.

2. Explain why:

   (a) metal atoms form positive ions;
   (b) non-metal atoms form negative ions.

3. Explain what the following words mean:

   (a) electrolysis
   (b) electrodes
   (c) anode
   (d) electrolyte
   (e) electrolytic cell

4. Draw a labelled diagram to show the electrolysis of copper sulphate solution. Use a 6 V battery and carbon electrode.

5. Four molten salts were electrolysed with the results listed below. Sadly the pupil did not write down the results when they occurred. Complete the table for him.

| Molten salt | Formed at anode | Formed at cathode |
|---|---|---|
| Sodium bromide | Bromine | |
| | Chlorine | Calcium |
| Lead iodide | | |
| | Iodine | Potassium |

6. Explain why solid sodium chloride does not conduct electricity but molten sodium chloride can be electrolysed.

7. Draw a labelled diagram to show how you could copper-plate a drawing pin.

## ... electricity in chemistry?

# What do you know about ...

8. You decide to make an electric cell using a lemon (containing citric acid) and strips of copper, iron, magnesium. You push the strips into the lemon and connect a voltmeter between them as shown below.

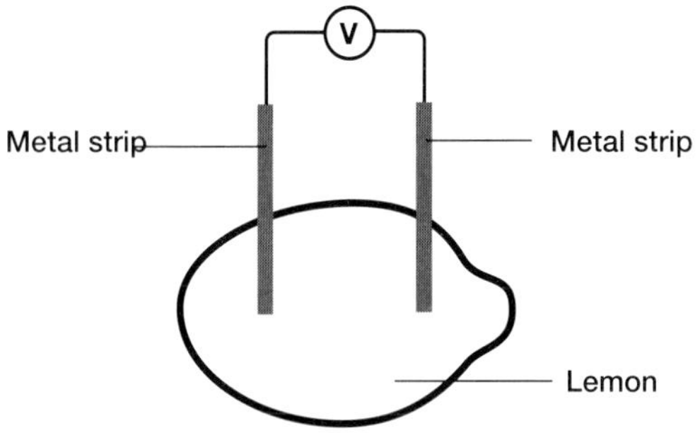


You try the following combinations of metals:

(a) copper – iron
(b) copper – magnesium
(c) iron – magnesium
(d) iron – iron

Which combination, giving your reason, will:

(a) produce the highest voltage reading?
(b) provide no voltage reading at all?

9. Your torch may well be powered by a 1.5 V battery; your family car may have a 12 V battery in it.

The word 'battery' is incorrectly used in one of the cases mentioned above.

Say why the word is incorrectly used in one of the examples but is quite correct in the other.

## ... electricity in chemistry?

# Further questions on ...

1. Sodium metal has a structure which is often described as a lattice of sodium ions floating in a 'sea' of free electrons.

   (a) Explain the following words:
      (i) lattice
      (ii) sodium ions
      (iii) 'sea' of free electrons.
   (b) Draw a diagram to show the structure of sodium metal.
   (c) Using your answers to (a) and (b), explain why sodium metal:
      (i) is flexible
      (ii) conducts heat
      (iii) conducts electricity

2. Consider the reaction:

   | |
   |---|
   | **POTASSIUM + IODINE ⟶ POTASSIUM CHLORIDE** |

   Potassium chloride is an ionic compound.

   Explain what is happening during this reaction. (You should mention electrons and ions.)

3. (a) Draw a labelled diagram to show the structure of sodium chloride.
   (b) Use your answer to (a) to help you to explain why solid sodium chloride does not conduct electricity, but sodium chloride dissolved in water does.

4. Draw a labelled diagram to show the electrolysis of sodium chloride solution.

5. Malachite is a mineral containing copper carbonate. Describe, giving word equations where possible, how you could obtain copper from malachite:

   (a) adding sulphuric acid to the malachite and electrolysing the resulting solution (draw a diagram of the electrolysis);
   (b) reducing copper oxide using carbon.

6. (a) Giving a reason, say what you would expect to find at the cathode if you electrolysed:
      (i) sodium sulphate solution
      (ii) copper sulphate solution
      (iii) lead sulphate solution

   (b) One of the results above is different from the other two. Why?

## ... electricity in chemistry

# Index

## G



## H



## I



## K



## L



## M



## N



## O



## P



## R



## S

## U



## Z